SS GREAT BRITAIN

1843–1937 onwards

First published in May 2012

Brian Lavery has asserted his moral right to be identified as the author of this work.

A catalogue record for this book is available from the British Library.

ISBN 978 0 85733 105 2

Library of Congress control no. 2011943923

Published by Haynes Publishing,
Sparkford, Yeovil, Somerset BA22 7JJ, UK
Tel: 01963 442030 Fax: 01963 440001
Int. tel: +44 1963 442030 Int. fax: +44 1963 440001
E-mail: sales@haynes.co.uk
Website: www.haynes.co.uk

Haynes North America Inc.,
861 Lawrence Drive, Newbury Park,
California 91320, USA.

Printed in the USA by Odcombe Press LP,
1299 Bridgestone Parkway, La Vergne, TN 37086.

COVER CUTAWAY:
SS Great Britain. *(John Lawson)*

SS GREAT BRITAIN

BRUNEL'S
SS GREAT BRITAIN™

1843–1937 onwards

Enthusiasts' Manual

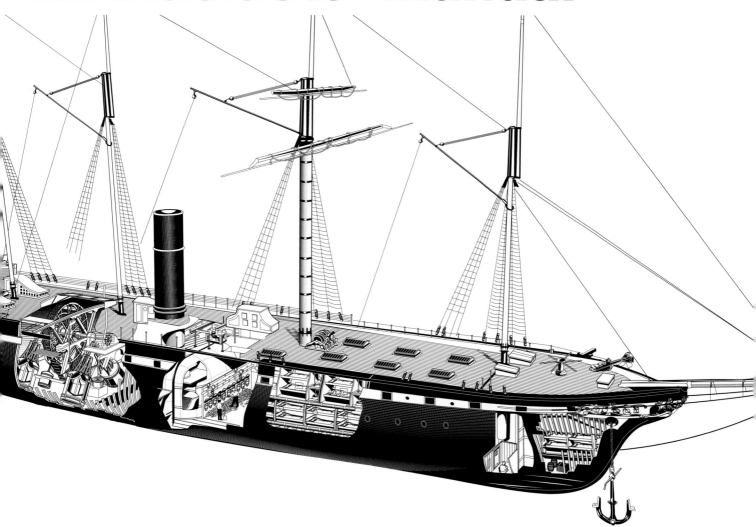

An insight into the design, construction and
operation of Brunel's famous passenger ship

Brian Lavery

SS GREAT BRITAIN

Contents

Introduction and acknowledgements

BELOW Brunel's SS *Great Britain* is the world's first great ocean liner. Launched in 1843 to provide luxury travel to New York, the iron-hulled steamship revolutionised travel and set new standards in engineering, reliability and speed. Today the *Great Britain* is one of the UK's most exciting museums and visitor attractions and in 2011 she welcomed more than 170,000 visitors. *(Mandy Reynolds)*

The Steam Ship *Great Britain* is a beautiful and historic vessel, which lies at the centre of Bristol life. She is one of the top four preserved ships in the United Kingdom, alongside HMS *Victory*, HMS *Warrior* and the *Cutty Sark*, and is probably among the ten most important historic ships in the world.

Her survival through a long exile in the Falkland Islands is purely a matter of chance, helped by the strength of her original construction. I think I first heard of her in 1960, in a little volume called quaintly *The Dumpy Pocket Book of Veterans of Road, Rail, Sea and Air*, which noted, 'She was beached at Sparrow Cove in 1937, where she remains today, some 114 years after she first put to sea.' Ten years later I was one of millions who watched enthralled on television as she was towed up the River Avon and under the Clifton Suspension Bridge, the result of the vision of Ewan Corlett and Richard Goold-Adams. Since then she has gone from being a 'rust-bucket' to one of Bristol's and the nation's greatest treasures.

Research into the *Great Britain* has been made easier by the fact that the great majority of the sources have already been collected, either as originals or photocopies, in the archives. It has also been helped by the cooperative attitude of the staff of the ship. In the *Great Britain*, particular thanks are due to Matthew Tanner, Rhiann Tritton, Eleni Papavasileiou, Joanna Thomas and Michael Conway. I am grateful also to the Trustees of the ss Great Britain Trust for their support and commitment to this project. More general thanks are due to the other members of staff who were always helpful. I have also used the resources of the National Maritime Museum, Merseyside

ABOVE The *Great Britain*'s glass 'sea' is proving to be the greatest wow factor for visitors. In true 'Brunelian' spirit, the ship appears to be afloat on what is the roof of a giant dehumidification chamber. Technological innovation, inspirational design and some truly revolutionary features were key to the ship's early success, and remain essential to her ongoing conservation in the 21st century. *(Mandy Reynolds)*

Maritime Museum, Bristol City Library, London Library and British Library and would like to thank the staffs of these institutions.

Thanks are also due to Tony and Marion Morrison for their kind permission to reproduce a small selection of the remarkable photographs they took of the *Great Britain* in the Falkland Islands during 1969–70, while making a television documentary of the salvage operation for the BBC; and TopFoto/Topham Picturepoint for permission to reproduce images from their collection. Photographs by Paul Blakemore, Alistair Hood, Pawel Libera, David Noton and Mandy Reynolds are by courtesy of the ss Great Britain Trust.

LEFT The process of conservation can never stop and costs more than £1 million per year. The ss Great Britain Trust receives no funding from central or local government and relies on visitors and events held on board as well as the generosity of members and supporters to fund work. *(David Noton)*

Chapter One

The *Great Britain* story

At the time of her launch in 1843, the SS *Great Britain* was the largest ship in the world. She was more than 100ft longer than her rivals and as the world's first iron-hulled, steam-powered ocean-going ship her design was revolutionary. Conceived initially for the transatlantic luxury passenger trade, she became successively an emigrant clipper, troopship, windjammer, and finally a coal hulk.

OPPOSITE The SS *Great Britain* moored in the Bristol Floating Harbour sometime after her launch. From a painting by Joseph Walter. *(By courtesy of the ss Great Britain Trust)*

ABOVE Isambard Kingdom Brunel (1806–59) was the designer of the SS *Great Britain* and the Great Western Railway among many other civil and marine engineering achievements. His innovative designs revolutionised modern engineering practice. *(Illustrated London News)*

RIGHT William Patterson the shipbuilder (1765–1869). *(Bristol City Museum)*

Conception

The Great Western Railway from London to Bristol had not quite begun construction in October 1835 when Isambard Kingdom Brunel, the young chief engineer, remarked half-jokingly that the line might be extended to New York by means of a steamship. The directors took him seriously and he built the *Great Western*, which was completed in 1837. She lost narrowly to the *Sirius* to become the first ship to cross the Atlantic under steam alone, but she was a much more suitable ship for the route and eventually made 67 voyages. The directors of the Great Western Steamship Company began to plan another ship to follow it.

Brunel was born in 1806, the son of an émigré French engineer who designed revolutionary block-making machinery for the Royal Navy. The young Isambard had helped his father with the first tunnel under the River Thames, but was seriously injured there in 1828. He convalesced in the Bristol suburb of Clifton and won a competition to design a suspension bridge over the Avon Gorge – though it was not completed until 1864, after his death. Brunel made the acquaintance of many Bristol businessmen, and improved the city docks before being appointed to the Great Western Railway in 1833 – at the time there were only 208 miles of railway track in Britain, mostly in the north of England. The small, dynamic man was also appointed consulting engineer to the Great Western Steamship Company, but his role was far wider than that and in many ways he was the main motivator. The boundaries between the different branches of engineering had not yet been drawn, so Brunel moved effortlessly between civil and mechanical, while taking in naval architecture in his stride.

The shipping world was changing rapidly after centuries of gradual evolution. The Americans were building much faster sailing ships, which would develop into the famous clippers. In Britain, the long-standing monopoly of the East India Company was abolished in 1834, while Parliament was moving towards free trade, which would open shipping up to foreign competition. Statistically the steamship was only a minor factor – in 1839 there were

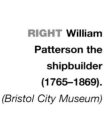

723 of them registered in Britain, of 80,000 tons, compared with nearly 21,000 sailing ships of 2.5 million tons. But steam power helped the shipping industry in many other ways – rail links to ports as exploited by Brunel, new channels dredged, breakwaters built and steam tugs to manoeuvre sailing ships into harbour.

The new ship was to be built in Bristol where the *Great Western* had been constructed and Brunel had a shipbuilder he could work with. William Patterson had been born in poverty in Scotland in 1795. He became apprenticed to a shipwright in Rotherhithe, London, and moved to Bristol in 1822 where he took over a yard in 1830. He had a great range of experience, from very early steamships to small but fast sailing ships of advanced design, even before the clipper ship became well known. Bristol had a long history as a port, but it was 6½ miles up the narrow, winding and highly tidal River Avon. In 1809 William Jessop had rectified some of the problems by sealing off and bypassing part of the river, to be known as the Floating

Harbour. This was accessible through the Cumberland Basin and a 44ft-wide lock.

The *Great Western* had been revolutionary in purpose and innovative in size, but apart from that she was a conventional design, with paddle wheels, a square hull section and wooden construction. The design of her successor started in September 1838 with a wooden paddle steamer 236ft long, 35ft broad and of 1,340 tons, only slightly bigger than the *Great Western*. But Brunel was never one to think small; this expanded to 2,000 tons by August, then in January 1839 the first truly radical decision was made – she was to be of 2,430 tons

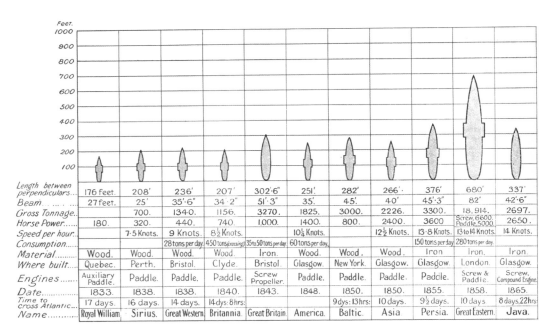

Length between perpendiculars...	176 feet.	208'	236'	207'	302'·6"	251'.	282'	266'·	376'	680'	337'
Beam...	27 feet.	25'	35'·6"	34'·2"	51'·3"	35'.	45'.	40'	45'·3"	82'	42'·6"
Gross Tonnage.		700.	1340.	1156.	3270.	1825.	3000.	2226.	3300.	18,914.	2697.
Horse Power......	180.	320.	440.	740.	1,000.	1400.	800.	2400.	3600.	Screw,6600 Paddle,5000.	2650.
Speed per hour.		7·5 Knots.	9 Knots.	8½ Knots.		10¼ Knots.		12½ Knots.	13·8 Knots.	13 to 14 Knots.	14 Knots.
Consumption......			28 tons per day.	450 tons (crossing)	35 to 50 tons per day.	60 tons per day.			150 tons per day.	280 tons per day.	
Material.........	Wood.	Wood.	Wood.	Wood.	Iron.	Wood.	Wood.	Wood.	Iron.	Iron.	Iron.
Where built....	Quebec.	Perth.	Bristol.	Clyde.	Bristol.	Glasgow.	New York.	Glasgow.	Glasgow.	London.	Glasgow.
Engines......	Auxiliary Paddle.	Paddle.	Paddle.	Paddle.	Screw Propeller.	Paddle.	Paddle.	Paddle.	Paddle.	Screw & Paddle.	Screw, Compound Engine.
Date.............	1833.	1838.	1838.	1840.	1843.	1848.		1850.	1855.	1858.	1865.
Time to cross Atlantic...	17 days.	16 days.	14 days.	14 dys. 8 hrs.			9 dys 13 hrs.	10 days.	9½ days.	10 days	8 days, 22 hrs.
Name............	Royal William.	Sirius.	Great Western.	Britannia.	Great Britain.	America.	Baltic.	Asia.	Persia.	Great Eastern.	Java.

RIGHT The place of the *Great Britain* in the growth of the Atlantic liner. Although she was a great advance in her day, the leap was not as great as with Brunel's *Great Eastern* of 1858.

BELOW A view over the River Avon of the Cumberland Basin, looking towards Bristol Harbour in 1841. This picture shows the locks leading into the docks from the tidal River Avon through which a ship built in Bristol would have to exit. Shipping can be seen moored in the docks, which are surrounded by warehouses. From an engraving after the painting by William Henry Bartlett (1809–54). *(Oxford Science Archive/TopFoto)*

and 295ft long, much larger than any ship ever built.

The design was controlled by a committee, including Brunel. Thomas Guppy was a long-standing member of Brunel's circle, a Bristol businessman born in 1797 who ran the family sugar-refining business until he took up engineering. Lieutenant Christopher Claxton joined the Royal Navy in 1804 and had seen plenty of action in the Napoleonic Wars. He was appointed Harbour Master of Bristol in 1834. He was known by the courtesy title of 'captain'

though he had not served at sea since 1819. Patterson was the adviser on naval architecture.

The committee was already thinking about the possibilities of iron construction by October 1838, when the *Rainbow* called in at Bristol. The quality and reliability of iron had greatly improved in the last few decades. The use of steam ashore had encouraged blacksmiths and boilermakers to develop watertight joints by riveting plates together. It did not take much to see that iron could keep water out as well as in, but at first there were sceptics. When the

steel barge *Vulcan* was launched into the Forth and Clyde Canal in 1819, the crowds on shore expected her to sink instantly. Three years later the *Aaron Manby* was the first iron vessel to take to the seas, sailing from London to France. An engineer like Brunel could see distinct advantages in iron. The planks of wooden ships were not joined to one another and the hull was not rigid (despite improvements introduced for the Navy by Sir Robert Seppings), so the hulls of longer ships tended to sag at the ends. Large pieces of timber were in short supply after years of wooden shipbuilding; the frames and internal planking of a wooden ship had to be feet thick, taking up a large amount of the space; and wooden ships, especially steamships, were highly vulnerable to fire.

The iron ship *Rainbow* had been built in Birkenhead for cross-channel service in 1837 and displaced 580 tons. The directors of the GWSS ordered 'the most searching enquiries' into the new material and Claxton and Patterson made several passages in her, including a voyage to Antwerp. Claxton later reported:

It became manifest that iron would afford greater strength, greater buoyancy and more capacity at less expense than wood. In capacity alone, for instance, the Great Britain [as the ship was now known] gains considerably more than 600 tons. To make this clear to everyone, it is necessary to suppose that the angle-iron or ribs, the shelves, & c. & c., rolled out and added to the plates forming her sides – when an average thickness of two feet of timber is represented by an average thickness of 2½ inches of iron, with far better ties, more compact framework, and far better strength altogether than wood can, under any circumstances, afford.

But iron ships were still relatively small and the new ship would be 5½ times the size of the *Rainbow*. Wooden hulls could be repaired by the ship's carpenter, iron ones needed much greater facilities. Iron would cause the compass to deviate, which could cause immense navigational difficulties. There was no way of coppering an iron hull to prevent it from being fouled by barnacles and weeds. And no

one knew what would happen to iron during long ocean voyages. But in March 1839 a shareholders' meeting was told that, 'After . . . the most ample investigation,' the directors 'had determined to build your next vessel in iron' and they were 'far advanced in their preparation for her construction.'

That spring another major decision was made. The breadth was expanded to 50ft, enough to have cabins on either side of a central saloon, with the obvious problem that the ship had to get out of the harbour through a 44ft lock. The directors may have believed that the harbour board would employ Brunel to enlarge the lock, but that did not happen for the moment. Instead, the shape of the hull was designed so that it could fit with very high water and a very light load, and the greatest breadth would be lifted above the sides of the lock. This explains the underwater shape, though it does not account for the 'tumble-home' or narrowing of the hull above water.

The directors also decided that a ship of this size should be built in a dry dock and floated out, rather than on a building slip to be launched into the water. They told the shareholders,

the greater security from risk afforded by floating-off so large a construction in preference to launching her, which also obviates the necessity of insurance, and the economy and expedition which this mode of building secures in her construction, and in placing on board her boilers and machinery, led your Directors to the determination of excavating a dock in your premises, on the bank of the Float[ing Harbour].

Meanwhile, another innovative ship visited Bristol. The idea of the screw propeller had been around since classical times, but in 1835 the Swede John Ericson demonstrated his contra-rotating design. He was disappointed with the response in Britain and decamped to America. In the same year Francis Pettit Smith tested a model in a farm pond just north of London, then built a launch that was demonstrated to great effect in the Paddington Canal near where Brunel was building his great railway terminal. He found backers to construct the *Archimedes* of 237 tons. Smith's greatest contribution was to put the propeller just forward of the rudder to assist in steering. After tests off Dover the *Archimedes* began a round-Britain voyage and soon convinced sceptics such as the engineering writer John Bourne: 'Heretofore engineers had been almost unanimous in the opinion that a screw would occasion a serious loss of power from the obliquity of the action, and the consequent dispersion of the water, and it was concluded, therefore, that it would be ineligible as a propeller. . . . But it was impossible to resist facts such as the performance of the *Archimedes* afforded.'

After the *Archimedes* arrived in Bristol the building committee used her for experiments with eight different types of screw. Brunel made a very detailed report for the directors of the GWSS, while Claxton outlined the advantages of the screw in a shorter form.

The most prominent points of superiority are:
1st. The facility afforded in carrying canvas, inclination or heeling not affecting the motive power of the propeller. . . . 2nd. It can only be in the highest seas that the screw even partially quits the water, and then only for a few seconds at rare intervals, while with paddles the hollow of the seas constantly leaves both wheels exposed. . . . 3rd. The breadth of beam going into docks and basins. . . . 4th. The diminished chances from collisions at sea, where the paddle wheels and houses constantly suffer.
5th. . . . The difference in resistance to the wind, the paddle-boxes and their appendages creating nearly one half of the whole resistance of the body. . . . 6th. The ease with which sail may be carried. . . . and by no means the least advantage in getting rid of the top weight of frame, shafts, wheels, &c., &c. . . .

When the decision to use the screw was announced in March 1841 the shareholders cheered, though two of them put the matter to a ballot. Brunel carried out his own experiments in the *Archimedes*, but he did not have as much time as he would have wished. His first propeller for the *Great Britain* was six-bladed and fan-shaped.

The designers approached some of the most experienced marine engineers of the day. Maudslay and Field of London were long-established partners of Brunel, but they were not interested until they saw the revised specification of April 1839. It was always going to be difficult to find a builder for such large engines, as new sets of tools would be needed. Humphreys and Hall produced a scheme for

BELOW Francis Pettit Smith's first design for a screw propeller in 1836, consisting of 'a sort of screw or worm, made to revolve rapidly under water, in a recess or open space formed in that part of the after part of the vessel commonly called the dead rising or dead wood of the run.'

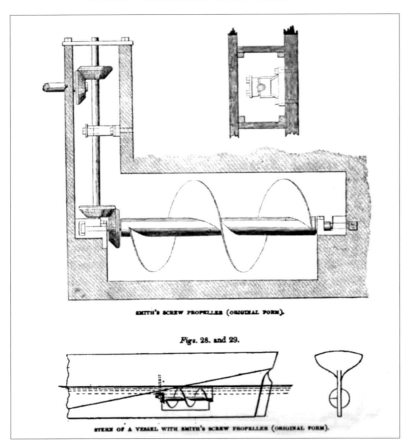

SMITH'S SCREW PROPELLER (ORIGINAL FORM).

Figs. 28. and 29.

STERN OF A VESSEL WITH SMITH'S SCREW PROPELLER (ORIGINAL FORM).

a trunk engine which had several advantages, but Brunel was very sceptical about their costings. Humphreys reported on 'the utter hopelessness of getting the engines made piecemeal' in Bristol and the directors felt they had no alternative than to make the engines themselves. Brunel was sceptical about this too. 'I think it is a course which must ultimately be adopted if the Company thrive – but I should have much preferred that it had been adopted gradually.' In March 1840 a dissenting shareholder wrote: '. . . Not content with being shipowners, they have taken upon themselves to become manufacturer of Iron Steam Vessels, and Steam Engines, upon a large scale. . . .'

When it was decided to use a screw propeller rather than paddles, the engine had to be moved through 90 degrees and the trunk type was no longer considered suitable. Apart from the need to use fuel economically on a long voyage, the greatest challenge for a marine engine designer was to fit it into the confined space on board ship. Brunel senior had already developed a four-cylinder 'triangle' engine in which pairs of cylinders were fixed to the bilges of the ship and met at a crankshaft several decks higher up. But the inexperience of the builders began to show – Claxton reported that the engines were 'somewhat on the patent of Sir Mark Brunel' and later Ewan Corlett assessed that '. . . in most other respects the original Triangle engine was considerably more sophisticated and ingenious' than that and 'the *Great Britain*'s boiler was a travesty of Sir Marc's fine effort'.

The triangular shape fitted quite well into the midship section. The 'double bottom' structure of the lower hull was modified to allow the cylinders to rest directly on the frame, and they were supported by massive diagonal beams reinforced with iron. Another factor was that a screw had to revolve at about three times the speed of a paddle. Brunel rejected the idea of gearing for this, for he had found the machinery on the *Archimedes* very noisy. Instead he conceived a system of chains.

Brunel and colleagues designed an unusual rig, intended to sail simultaneously with the engine running in order to maximise efficiency of both, as well as serve as a back-up propulsion system for each other as required. She was fitted with six masts, a daring plan when few

ships had more than three. The large number of masts obviously took advantage of her great length, and aided handling. As Claxton put it:

Nothing is so difficult to handle, under a variety of circumstances, as the sails of a steamer, unless the engine be stopped, which can never be allowed in Atlantic steaming, where onwards, and for ever onwards, is the rule. The greater the number of masts, the more handy the sails, and the smaller the number of seamen required to handle them.

The sails were mostly to be of the 'fore-and-aft' type, which created much less drag when not in use and could be handled by a smaller crew. Such a rig could sail closer to the wind, which would be more often required when working simultaneously with the steam engine since the apparent wind is often from ahead of the vessel. One mast, the second from the bows, was fitted with square rig in a more or less conventional manner. This could be useful if there was a strong wind directly behind. And it might appease the more conventional members of the design committee, and the sailors who would have to operate the ship. Aft of the second mast the masts declined gradually in size, so that the aftermost one was less than three-quarters of the height of the second one. They were tilted backwards, or raked, at increasing angles so that the last one was set at 10 degrees to the vertical.

The ship had come a long way since its first conception as a partner to the *Great Western*. According to Brunel's son (also Isambard), the committee had undertaken:

To design and construct a steam-ship larger than any that had, up to that time, been launched, to make this ship of a material which had lately been introduced into shipbuilding, and which had never before been employed on a large scale, to adapt to this ship a novel form of propeller which had not previously been used save in a merely experimental steamer, and to build in a newly opened manufactory marine engines of a much greater size than any that had hitherhto been contemplated, and of a totally different character, was indeed a bold enterprise.

The launch

When the launch day of the *Great Britain* finally came on 19 July 1843, it had its share of farce, as reported by the *Illustrated London News*. The ceremony was to be performed by Prince Albert, who arrived on the new Great Western Railway and was fêted through the 'irregular old streets' of Bristol. Six hundred people attended a banquet in the highly decorated pattern-makers' shop of the yard, while water was let into the dock to make the ship float for the first time.

The dock gates were open as His Royal Highness ascended to a gallery on the northern side of the dock, and the river ferry *Avon* began to haul the *Great Britain* out, watched by 30,000 people. Bands played and crowds cheered as the ship began to move. Mrs Miles, the wife one of the directors, swung a bottle of wine towards the bows, but it fell ten feet short. The tow rope broke and the ship began to swing towards the royal platform. Albert tried himself and the bottle broke over the heads of some of the men who were trying to fend the ship off from the dock. At last the great ship was hauled out at 3:15 in the afternoon. Many

of the crowd had a feeling of disappointment – a floating-out did not have the drama and spectacle of a traditional launch.

Getting out of dock

She spent more than a year in the Floating Harbour, fitting out and waiting for modifications to the dock and for a tide that was high enough to get her out. She was visited by the King of Prussia and photographed by the pioneer Fox Talbot – possibly the first photograph of a ship ever taken. Eventually the inner entrance to the Cumberland Basin was

modified 'in a most satisfactory manner' and the ship entered at the end of October 1844.

On 10 December the tide was reckoned to be high enough to get her through the outer lock into the River Avon and a tug was hired. She was about three-quarters of the way though when Claxton, on board the tug, noticed that she was stuck fast. It was in the nick of time, for the water would have soon started to fall and the vessel might have suffered severe damage with one end supported by the water and the other in the air. She was hauled quickly back into the basin. Brunel cancelled an urgent engagement in Wales and took charge of the

FAR LEFT The *Great Britain* floats out of the dock. *(Illustrated London News)*

LEFT The banquet in the pattern-makers' shop, wrongly described in some accounts as being on board the ship. *(Illustrated London News)*

BELOW Cumberland Basin in the early 20th century. The North Entrance Lock, to the left, had been much enlarged by that time. *(Jonathan Falconer collection)*

work to modify the dock temporarily and get the ship out the next day, before the tide levels began to decline.

I have been hard at work all day altering the masonry of the lock. Tonight, on our last tide, we have succeeded in getting her through but being dark we have been obliged to ground her outside. I confess I cannot leave until I see her afloat again and all clear of her difficulties.

William Prideaux was one of the local citizens who breathed a sigh of relief – 'The Great Britain is at last got out of Dock, a nice laugh against the Bristol people for building a vessel they could not get out of Dock.'

Trials

At 8:00 on the morning of 12 December the ship was towed down the Avon to the anchorage at King's Road in the Bristol Channel. Her boilers were filled with water and the engines started to run for a few hours at a maximum speed. Her trials were to begin properly on 8 January when she took on board 'a numerous party of proprietors, and several engineers and scientific men'. But the morning fog did not clear and the pilot was reluctant to start until he was persuaded to go a short distance, 'merely to gratify the visitors'. Nevertheless she made the high speed of 11½ knots. At last, on 20 January she had a proper run, 'down the Bristol Channel to Ilfracombe and back, a distance of 95 miles, without much wind, but in a heavy swell'. This was done in 8 hours and 43 minutes, with an average speed of above 11 knots.

On the evening of 23 January she was ready to leave the Bristol area for the last time in her working life to sail for London, despite 'the masses of cloud which had traversed the sky during the day, and the occasional heavy gusts of wind . . .'. A party of more than two dozen merchant and Royal Navy captains, civil engineers and doctors was on board. The anchor

was raised at seven but some wreckage from another ship had been caught in it and had to be cleared. The engines, according to Captain Crispin of the Royal Yacht, had only been used for about 15 hours up to then, so they had to be stopped occasionally as bearings overheated. Then the weather became even rougher. 'At 5 A.M. on the 24th the wind which hitherto had been from the S.W. blowing strong, suddenly chopped round to the N.W. and blew very hard. At 9 when off Lundy, we found a very heavy nasty cross sea, occasioned by a spring Ebb Tide running against the gale.' But by all accounts the ship coped with this remarkably well and Captain Crispin was full of praise.

I had therefore a fine opportunity of judging, both the qualities of the Ship and screw. The former I found to have all the quickness of Motion of a Vessel of 500 tons, more particularly in regard to her rolling. . . . At the same time I must observe, that her very wonderful stability when pressed with canvas was astonishing! – Her pitching was very easy, and although so light I watched most particularly the keel of the Ship from over the Stern, to see how far she rose and lifted the Screw out of the Water, and to my great surprise she never on any one occasion, had less than 6 feet of it immersed.

The ship suffered some damage to her bows, and 11 out of 12 engineers and most of the stokers were laid low by seasickness. It was a great relief when she passed the Longships rocks off Land's End at 8:45 in the morning, to find a flat and windless sea. She steamed up the English Channel and at 12:45 the next day she passed though the Needles Channel against an ebb tide, then anchored briefly off Cowes on the Isle of Wight to pick up dispatches. Thirteen hours later she was in the Downs off the east coast of Kent, and after a short stop she headed up the River Thames. There was satisfaction as she overtook the paddle steamer *Waterwitch*, and trepidation as she faced another gale. Off Gravesend she had to weave in and out of the numerous ships at anchor, but again her steering qualities were proved, despite her great length. As she headed upriver the crews of each ship came on deck to witness her, and she arrived at Blackwall where she dropped anchor at 3:30 in the afternoon of Sunday 26 January.

Exhibited in London

The directors of the GWSS were still in no hurry to get the ship into service, and she spent the next five months at Blackwall and open to the public, at a cost. The most

BELOW Crowds cheer as Queen Victoria visits the *Great Britain* on the Thames at Blackwall. *(Illustrated London News)*

distinguished of her visitors was Queen Victoria, who wrote in her diary:

> . . . set off for Greenwich to visit the 'Great Britain' . . . we stepped from the 'Dwarf' on to a platform, to which accommodation stairs had been fixed, to enable us to get on board the 'Great Britain', which is an enormous ship. . . . The berths are very small, but there are nearly 300 of them. The contrivances and securities against danger are extremely ingenious. All the fittings & decorations are very fine & everything made as comfortable as possible, but I would not care to go to sea with such a quantity of passengers, or in such a long narrow ship.

On 12 June the ship set sail for Liverpool, which was to be her home port for the rest of her active life. If anything, she attracted more attention than on her first voyage, with crowds lining the banks of the Thames, and even the hilltops of the coastal areas which she passed. She stopped at Plymouth to take a party out to the Eddystone Lighthouse. She visited Dublin and arrived in Liverpool where she was opened to the public in Coburg Dock, attracting up to 2,500 visitors a day. Finally in July 1845, more than two years after her launch and eight after her conception, she set sail for New York.

BELOW The ship returns to Liverpool after the first screw-propelled crossing of the Atlantic.
(ss GB Trust)

New York at last

Queen Victoria's fear of '. . . such a quantity of passengers' was not proving to be a problem – only 45 berths had been sold for the voyage. She left Liverpool on the 26th to the kind of fanfare that her crew must have become used to, with crowds lining the banks and small boats in the river. She turned to sail round the north of Ireland, but met bad weather as soon as she was clear of the coast, with gale force winds from the north and a great sea in front of her, to the horror of her crew who 'showed great unwillingness to do their duty this day'. The weather moderated and she made good progress under steam at 9 knots.

Off Newfoundland Captain Hosken did not slow down, though the vague shapes of ships could often be seen through the notorious fogs. Her leadsmen began to find the sea bottom off Nantucket on 8 August and she entered New York Harbour two days later after a passage of 14 days and 21 hours. She had a welcome there, too, and the New York Herald reported that she was 'truly beautiful' and 'seemed a miniature city'. Yet again she was opened to the public, though the Brooklyn Eagle claimed that prices were too high and 'there were comparatively few on board yesterday – the curiosity of our citizens being much less than was anticipated'.

Great Britain set sail on 30 August, after some repairs in dry dock, to an even greater send-off but carrying only 53 passengers. She began her second crossing from Liverpool on 27 September with a more promising figure of 102 passengers and soon ran into trouble. After several mishaps and a grounding off Nantucket, a passenger wrote from anchor off Martha's Vineyard,

> Here we are after a voyage so far which has been tedious and not without danger. Our fuel nearly exhausted although we are still 200 miles from New York. Our foremast is gone, three arms of the propeller [though] I have not been able to detect any irregularity in the engines. It appears that after Saturday a strong current carried us right among the breakers of Nantucket Island called the Rips.

When she eventually reached New York she

was put in dry dock and it was indeed found that two of the blades were gone, so one was shifted to the other side to make it more symmetrical. But on the voyage back one of them began to strike the sternpost then broke off. Another broke the next day, leaving only one complete, but the ship behaved well under sail and steam until the last arm broke and she had to do the final part under sail. When she overtook two other sailing ships, Hosken wrote in his log, 'This is wonderful with our little spread of canvas and more than I expected well as I thought of her sailing qualities.' She berthed in the Mersey with the aid of tugs.

Great Britain was taken into dock for the winter and some of her more adventurous features were removed – the wire rigging was replaced with ordinary hemp, the mast behind the funnel was removed and what was now the third mast went over to square rig. A spare, four-bladed propeller was taken out of store and improvements were made to the steam pumps of the engine and fire flues of the boilers, all of which seemed to improve her performance

ABOVE Aground in Dundrum Bay, Ireland, in September 1846. The tide is out, exposing the hull, and the Mountains of Mourne are in the background. *(Illustrated London News)*

during trials in the Irish Sea in the spring of 1846. But her passenger list was even shorter at 28 when she crossed the Atlantic again in May, and one of her new air pumps broke down so she had to do the last six days under sail. On the way back she carried 42 and made a fast and relatively uneventful crossing of 13½ days. She left in May 1846 for her fourth double crossing with the more promising figure of 110 passengers, but grounded again when she lost her bearings off Newfoundland and scraped on Cape Broil Reef – though inspection back in Liverpool showed that she had not been damaged.

Dundrum Bay

Perversely the events of the last trips did not seem to have discredited the *Great Britain*, but proved that she could take punishment and perform well under sail. As a result, the next crossing was taken up by 180 passengers, almost enough to put the voyage into profit.

But that was not to be. After leaving Liverpool at 11:00 on 22 September, the ship sailed past the Calf of Man light, where she should have turned north, and mistakenly headed for the new St John's Bay light on the coast of Ireland. The mistake was discovered too late and at 9:30 that night she ran aground heavily in Dundrum Bay, to the distress of the passengers.

I had just gone into my stateroom and the instant I felt the shock I knew something was the matter. In a moment there was a second shock and all was confusion. Men and women rushed out, the latter from their berths, and some threw themselves into the arms of strangers. We could with difficulty stand. Oh I cannot tell you the anguish of that night.

The passengers were all saved but the ship was stuck fast. Hosken made the excuse that he had confused the two lighthouses because a chart he had bought in May did not show St John's, and he was backed up by the

RIGHT A frame was built under her stern and the ship was hauled out by a large team of labourers. *(Illustrated London News)*

GWSS board. But that was specious, he should have made sure that the charts were kept up to date and he should have had the route tracked by dead reckoning. The *Mechanics Magazine* was more frank in putting it down to 'the most egregious blundering'.

On the positive side, the ship had survived a jolt that would almost certainly have destroyed a wooden vessel and the concept of iron was proved to British shipbuilders. Brunel had 'very mixed feelings of satisfaction and pain almost mounting to anger'. Claxton took charge of operations, but attempts to pull her off by tug were frustrated by bad weather. She was holed slightly in the attempt and had to be driven further inshore to prevent her sinking.

Finding ideas for salvaging her almost became a national sport as dozens of suggestions flooded in, and Brunel was obliged to have a circular letter prepared declining assistance. He arrived to take charge early in December and he urged Claxton to build ever-higher breakwaters in the soft sand. 'If a six-bundle fagot won't reach out of the water, try a 20 bundle one. If hundredweights won't keep it down, try tons.' An elaborate wooden framework was constructed round the stern and the ship was lightened by taking out coal and fittings.

In May 1847, Brunel advised that a salvage expert, James Bremner, be employed. The task now was to get the ship out of the hole she had dug for herself in the sand and Bremner used ideas that seemed almost as crazy as the unsolicited ones sent to Brunel. A giant lever was used to raise the bows, with wedges driven under the keel, while an ingenious arrangement of chutes and rammers forced pebbles under the hull. The holes in the hull were sealed up and hundreds of labourers dug a trench in the sand towards the sea. Wooden boxes known as camels were fixed under the hull to give extra buoyancy. Ireland was in the throes of the Great Famine at the time and the work must have provided welcome relief for a few at least of her people.

The first attempt to float the *Great Britain* on 13 July failed, but on 29 July she was dragged out towards the *Birkenhead* anchored offshore. Claxton wrote to Brunel, 'Huzza! Huzza!' but the troubles were not over. The tow rope broke on

the way to Liverpool and was quickly replaced. She was put on the gridiron in Princes Dock and inspected. She was found to be 'safe and sound in hull and frames, not bent, shaken or strained . . .'.

All this meant the end of the Great Western Steamship Company and its main remaining asset, the *Great Western*, was sold. The *Great Britain* had cost £120,000 to build, she was insured for only £17,000, and the sale of the hull fetched £20,000, half of what had been hoped for.

Gibbs, Bright and the great refit

The *Great Britain*'s new owner was Gibbs, Bright & Co. of Bristol and Liverpool, which operated the Eagle Line of sailing ships to Australia. The company was no stranger to the *Great Britain* as it had been her agent; and before that, it had played a part in the formation of the Great Western Railway. At the same time,

ABOVE One of the rocks which the *Great Britain* closely missed in Dundrum Bay. *(Author)*

BELOW The new four-masted rig, with twin funnels. *(ss GB Trust)*

news arrived that gold had been discovered in Australia. The British colonies there had been established 60 years earlier as dumping grounds for convicts and had developed slowly. Tens of thousands of adventurers from Britain and Europe were likely to follow the example of the Californian Gold Rush two years earlier, and were expected to pay high prices to reach Australia quickly. The *Great Britain* as a fast, long-range steam and sailing ship was well equipped to transport them.

Repair and conversion was supervised by William Patterson and included a new engine built by the leading manufacturers of the day, John Penn & Sons of Greenwich. It was on the oscillating principle and took up less space and was more efficient than the original. The ship now had two funnels, side by side. The cabin accommodation was reorganised to provide for large numbers of steerage passengers and a deckhouse was added in the stern for saloon passengers. Another new rig was fitted, with two square-rigged masts before and aft of the funnel, and fore-and-aft rigged ones in the bow and stern.

Despite the rush to get to Australia, the owners decided on a trial voyage to New York under her new captain, Barnard Mathews. She left Liverpool on 1 May 1852 carrying 180 passengers, ironically the same number as five years before when she was nearly wrecked.

She was back on Merseyside a month later to prepare for her longest voyage yet.

The first Australian voyage

The *Great Britain* sailed from Liverpool on 21 August carrying 630 passengers, a number that the GWSS could only have dreamed of. Captain Matthews was treating her as a steamship rather than sail-assisted, for he ignored the sailing ship route to go against the current and prevailing winds to refuel at Cape Town. But a few hundred miles from the Cape, he was obliged to call the passengers together and tell them '. . . that from the continued state of the adverse winds it has been thought fit to return to St Helena for coals'. He blamed the engineers, claiming, 'I have been misled as to the amount of coal on board.'

Neither crew nor passengers were happy and gold prospectors were not the most refined or patient of people. Seaman John McFall wrote, 'They arm themselves every night with pistols and dirks so if there is a row who knows what to expect.' The crew was little better – seamen had to be confined and gagged, and on one occasion the meal was late because the captain was breaking a fight among the cooks.

After taking on very expensive coal on the remote island, the ship went on its way and took on more fuel at Cape Town. She arrived in

RIGHT The *Great Britain* leaving the Mersey in four-mast two-funnel rig. Perch Rock Lighthouse and the New Brighton fort can be seen behind her stern. She is possibly on her first voyage to New York on 1 May 1852. *(Illustrated London News)*

Melbourne after a voyage of 81 days, about 20 longer than expected. In future she would follow the sailing ship route to Australia.

Getting into her stride

Early in 1857 the ship had been taken into dock for another refit. Her engines were satisfactory after some teething troubles, but her two funnels were changed into one. The *Great Britain* was now in her most successful phase. She had different engines, a different rig, more decks, a different route and a different command style from her original plan, but she justified the concept of a large iron screw steamer and she turned a profit. She made 28 round trips to Australia between 1858 and 1876 and carried a total of about 15,000 passengers out and 10,000 back. She gained a reputation as a reliable, safe and fast ship, with a happy atmosphere on board under Captain John Gray.

The quality of her emigrants had probably improved; they were no longer adventurers like the gold-diggers of 1852, but respectable artisans and labourers trying to make a better life in Australia. It was during this period that life and death on board were recorded in some detail in letters, diaries and ship's newspapers. All classes were represented, reflecting increasing literacy among the passengers.

A sailing ship

By 1876, much had changed in the shipping world in the more than 30 years since the *Great Britain* was launched. The paddle wheel was obsolete except in the shallow waters of rivers and estuaries. Iron ships were now common and 72,400 tons of iron shipping were built in Britain in 1881 compared with 16,500 tons of wood. Even more striking was the improvement in steam power. The compound engine used higher pressure and was replaced by the triple expansion engine, first used at sea in 1876. It was far more fuel efficient than the engines of the *Great Britain* and non-stop voyages to Australia were more practical. Great shipping lines like White Star, P&O and Alfred Holt began to dominate the seas.

The qualities of her original design enabled her to find a new life as a sailing ship. After failing to reach her reserve price at auction in 1881, she was taken up by Anthony Gibbs of London,

The *Great Britain* at war

In March 1854, while the *Great Britain* was preparing for her third Australian voyage, Britain became involved in the Crimean War. The ship was allowed to proceed, though on the voyage there were alarms about Russian raiders. She was put into quarantine in Melbourne due to a death from smallpox, and to celebrate her release Captain Gray fired off his six cannons in a 'demonstration of joy and congratulation'. This was taken up by other ships in the harbour and the resulting noise caused the citizens ashore to believe that the Russian Pacific Fleet was invading, and mobilised the army and militia. After the dust had settled, 'They were by no means pleased . . . at the Captain of the *Great Britain*. . . .'

On her return to Liverpool the ship was more directly involved in the conflict when she was hired to transport troops to the war zone. She could carry up to 1,650 officers and men

and 30 horses. Picking up troops at Cork, the common stereotypes of the army of the day were soon reinforced. The officers were arrogant and one said he 'did not know where his men were stowed'. Of the troops themselves, Heywood Bright of the shipping company wrote, 'A nastier, dirtier, more stupid set of men I never saw.' After dropping them off at Malta the ship brought back invalids, probably including survivors of the Battle of Balaclava and the Charge of the Light Brigade.

On another trip to Malta she took on French troops who were 'much quicker, are always merry and never grumble . . . except for their dirt they beat the English all to pieces'. They were delivered to the improvised harbour of Kamiesch in the Crimea, then the ship took a party to the recently captured town of Kertch on the other side of the peninsula. When the allies attacked the Malakof Redoubt at Sevastopol in June 1855, the *Great Britain* was not far away. Heywood Bright reported, '. . . the fireing certainly was very heavy when we arrived, and it is nearly as bad now'.

After bringing troops back from the war, the ship was refitted in Liverpool yet again. The twin funnels were replaced by a single one and the deckhouse was extended to form a complete deck. Her three new masts were all square rigged and a lifting screw was fitted, signalling that she was finally recognised as a sailing ship assisted by steam. This rig was particularly suited for long voyages running before the wind in the Roaring Forties. Another round trip to Australia took place between February and August 1857, before the *Great Britain* was caught up in another conflict. In May 1857 the sepoys of Britain's Indian Army mutinied and the ship was chartered to take out reinforcements from the 17th Lancer Regiment and the 8th Hussars. She left Queenstown (now Cobh) in Ireland on 6 October, while the band played 'Cheer, Boys, Cheer'. At Cape Town they heard that Delhi and Lucknow had been relieved – though the affair was far from over when the *Great Britain* arrived in Bombay in December.

which had absorbed Gibbs, Bright. Her engine and passenger accommodation were removed to create extra cargo space. Her foremast was moved forward to balance the rig. Her hull was covered with wood sheathing, to prevent fouling, to strengthen the aged iron plates or to protect them from bumping. She was to trade between Liverpool and San Francisco.

Once loaded with a cargo of nearly 3,300 tons of coal for the outward voyage, the maximum breadth in the French frigate-type midship section would be underwater, and her crew refused to sail unless she was lightened. As soon as the tug was cast off near Arklow, Captain Stap found 'the caper the ship cut that night is almost indescribable'. The crew ('a more useless lot I never was with') threatened mutiny and he was forced to put into Queenstown for shelter. On the way south he lost sails in storms and found there was no condenser to provide a supply of water – 'I almost wish now that I had not taken the command'; 19 of his men deserted at Montevideo. He reached San Francisco in June 1883 after six months, and was back in Liverpool by August.

Stap remained in command for a further round trip in 1884–85, before setting out again in February 1886. Off Cape Horn she met strong gales. The cargo shifted, masts were lost and he was obliged to put into Port Stanley in the Falkland Islands, where she would be used as a hulk for storing wool.

The *Great Britain* was now a curiosity and unable to be insured at A1 status at Lloyd's because she did not conform to the contemporary rules for iron shipbuilding.

BELOW The ship laid up in the Falkland Islands, with her lower masts still intact. *(ss GB Trust)*

Chapter Two

Anatomy of the *Great Britain*

'Her lines are very beautiful and adapted to attain the highest rate of speed. Her exquisite proportions prevent that appearance of heaviness, which is generally observable in large ships; and, although she is probably the strongest vessel ever built, she has a remarkable air of grace and lightness.'

(John Grantham, engineer, 1842)

OPPOSITE The five-blade propeller and rudder seen beneath the 'waterline' of the humidity-controlled dry dock. *(Mandy Reynolds)*

The shape of the hull

A visitor today can walk round the *Great Britain* in her dry dock in Bristol and examine her hull both below and above water – a shape which Christopher Claxton described as 'somewhat peculiar, yet accordant with the taste (when she was built) of many nautical men'.

William Patterson drew the lines of the ship, but the other members of the design committee had a great deal of input, as well as outside experts. A naval architect would be given the basic dimensions of the hull, including length, breadth and depth. Once they were settled the designer drew the profile, or side view, of the hull. The basic shape was determined by the number of decks, five in this case. The hull was of course very long, and the fashion of the time made it quite low-sided. The high castles of past ages at the bow and stern had been gradually reduced over the centuries and had almost disappeared in the more peaceful world after 1815. A low hull meant the ship was less likely to be driven sideways by the wind, and under steam it created less air resistance against relatively weak engines. Ship designers had not yet acquired the habit of building superstructure above the decks.

Earlier shipbuilders had favoured a large amount of sheer, in which the bow and stern curved upwards to meet the waves. This too was out of fashion, the decks and bulwarks of the *Great Britain* rose only slightly in very gentle curves. At the bow, the stem post provided an echo of the beakheads that had dominated large ships in past ages, but with far simpler construction. The stern had a radical design below the water, with space cut out for the screw propeller, and a new type of 'balanced rudder'. Above the waterline it too followed a much more traditional shape, with a flat rear face angled over the water to provide windows.

The designer went on to draw the midship section, the widest part of the ship. This was a rather surprising shape in the *Great Britain* with an almost flat section known as the floor extending 12ft on either side of the keel, a bilge in which the shape turned to rise about 30 degrees from the vertical to the widest point or 'maximum breadth' about 25ft above that and about 5ft above the waterline. After that the shape turned inwards, a practice known as tumblehome. The flat floor and the sharp 'turn of the bilge' can be explained by the need to get the ship through the Cumberland Basin on completion, but the tumblehome is less easy to account for. It had been common in warships in the 18th century, but modern warships had little

BELOW The lines of the *Great Britain*. The sheer plan is to the right, with vertical lines showing the main frames, and horizontal waterlines. The curved lines are the buttock lines constructed from the verticals on the body plan below. The body plan shows the frames at each station, seen from aft on the left-hand side, and from forward on the right. The waterlines are shown in the half-breadth plan below the sheer plan. *(Bourne, Treatise on the Screw Propeller)*

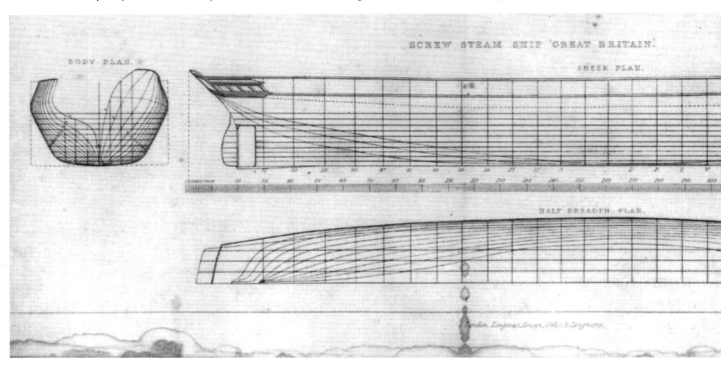

if any. It reduced the width of the upper hull, which might create instability if the ship heeled so that the maximum breadth was underwater.

Brunel had experimented with ship design with the help of his father, a former French naval officer. The six-sided midship section had been common in French frigates and corvettes up to that time, and they were associated with high speed, especially when captured by the British. Brunel senior might have absorbed this in his work in the British Royal Dockyards and adopted the archaic midship section in its entirety. Christopher Claxton thought that the tumblehome gave 'a man-of-war like appearance and wholesome rotundity in the after-body'.

It was normal to draw the midship section by means of a series of segments of circles, and this seems to have been done in the *Great Britain*. In conventional practice, the design was then gradually altered to the ideal shape for the bow and stern by means of a series of construction lines known as rising and narrowing lines. After that the shipwright drew horizontal 'waterlines' and vertical 'buttock lines' to test if the shape was streamlined or 'fair'. But Patterson was an advanced designer. In 1834 he had built the schooner *Velox*, which had been noted in the Bristol press as 'an improvement

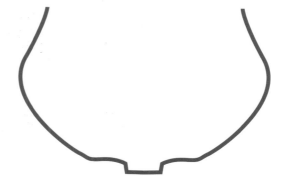

in the style of nautical architecture [which] must be hailed by everyone with satisfaction'. For him, the shape of the waterlines and buttock lines controlled the whole procedure. Fast sailing ships were usually given concave waterlines near the bow and stern, in contrast to the bluff bows of previous generations. A straight buttock line in the stern was regarded as another factor in creating speed.

British designers used a combination of plans and models, according to John Grantham:

> . . . the draftsman lays out the lines of the vessel. . . . From this it is usual to prepare a model, to be constructed as follows: Thin slabs of wood, equal in thickness to the horizontal or water lines of the drawing, are carefully prepared, and to these are

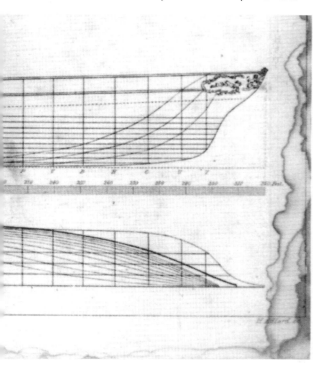

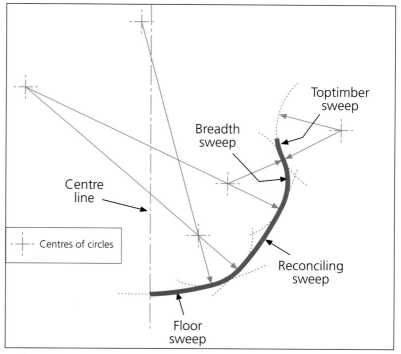

Toptimber sweep

Breadth sweep

Centre line

Centres of circles

Reconciling sweep

Floor sweep

transferred the half-breadth lines of the vessel. Each board is cut to the shape of the water line thus drawn upon it; and the whole is either bolted or screwed together, and carefully pared down by a spokeshave. By this half model the eye is assisted to form a more correct judgement of the intended ship, and improvements can be made by the further use of the modeller's tools. . . . No-one who can appreciate the extreme quickness of the practised eye to detect errors in form, will despise this simple method.

The shareholders of the GWSS were told: 'The knowledge of various parties was also put in requisition, embodied in models; and after many modifications the Directors finally determined on the proportions shewn in the model now laid before you.' The fine 'entry' at the bows caused Christopher Claxton to claim 'The *Great Britain* sends no wave before her' in contrast to the *Rattler,* which Brunel built for the Navy and which 'throws a five or six feet wave before her' – steamships should not be 'retarded by a mountain wave generated by themselves'.

Above water the design of the upper bows was just as radical. The most striking element is the flared bow in which the lines curve sharply

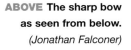

ABOVE The sharp bow as seen from below.
(Jonathan Falconer)

RIGHT The bow above the waterline, showing the pronounced flare.
(Jonathan Falconer)

outwards as they rise from the water. These would create a righting motion when the bows were buried in the waves, and would deflect spray outwards. In 1846 an American writer noticed a similarity between the hull form of the new clipper *Sea Witch* and the *Great Britain*. It is not likely that he knew much about the underwater hull of the British ship, but one common feature was the flared bow. Probably both came from the 'Aberdeen bow' first used by *Scottish Maid* in 1839. But not everyone liked it, and Commander William Crispin of the Royal Yacht *Victoria and Albert* commented after the *Great Britain*'s first voyage:

> *. . . were it not for the extraordinary flaring bow of this vessel, I should have no hesitation in pronouncing her as good a sea boat as I ever saw, this no doubt a great error which has proved by the force with which the sea struck her, so great was the shock that it lifted the deck of the forecastle and sprung and carried away a great many of the iron fastenings; however the directors see the defect, and assured me they would have the hollow part above the water all filled up, which can be very easily accomplished.*

This never happened until the ship was sheathed in wood as a sailing ship in 1882.

Above water at both the bow and the stern, the *Great Britain* kept up much of the appearance of traditional ship design, but little of the reality. The stern had a relatively flat surface with windows for the passengers. The bows did not use the traditional structure of beakhead, knee of the head and rails of the head. Instead, the whole structure was a single item, integral with the main hull, but the overall shape was rather similar, with a double-curved cutwater under the figurehead.

The design committees had produced a very fine hull shape, apart from the anachronistic tumblehome. John Grantham wrote in 1842: '. . . her lines are very beautiful, and adapted to attain the highest rate of speed. . . . Her exquisite proportions prevent that appearance of heaviness which is generally observable in large ships; and, although she is probably the strongest vessel ever built, she has a

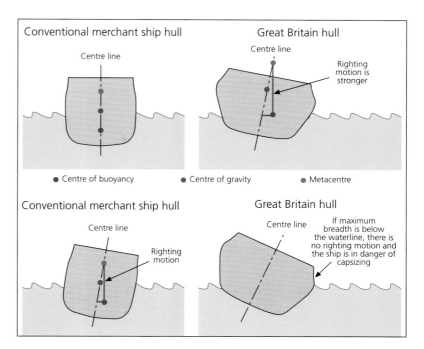

remarkable air of grace and lightness.' The designers had combined the two classic fast ships of the age, the French frigate and the clipper. Her lines gave her a high metacentre, or centre of flotation, which meant that she tended to roll excessively when there was no sail to steady her. As Allan Gilmour wrote in 1852: '. . . At the present moment the ship is rolling so much that things are falling off the table. . . . The piano in the music room positively shifted at least a foot.' On the other hand her great length gave her some advantage in rough weather: '. . . The length of 350 feet [sic] is sufficient to reach over three of the ordinary waves of the Atlantic, which supports her continually at several parts of the length of the hull. . . .'

ABOVE The metacentre and its effect on stability. *(Matthew Marke)*

BELOW The effect of a hollow bow on stability. *(Matthew Marke)*

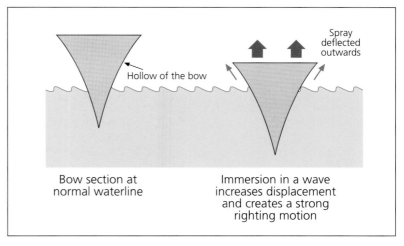

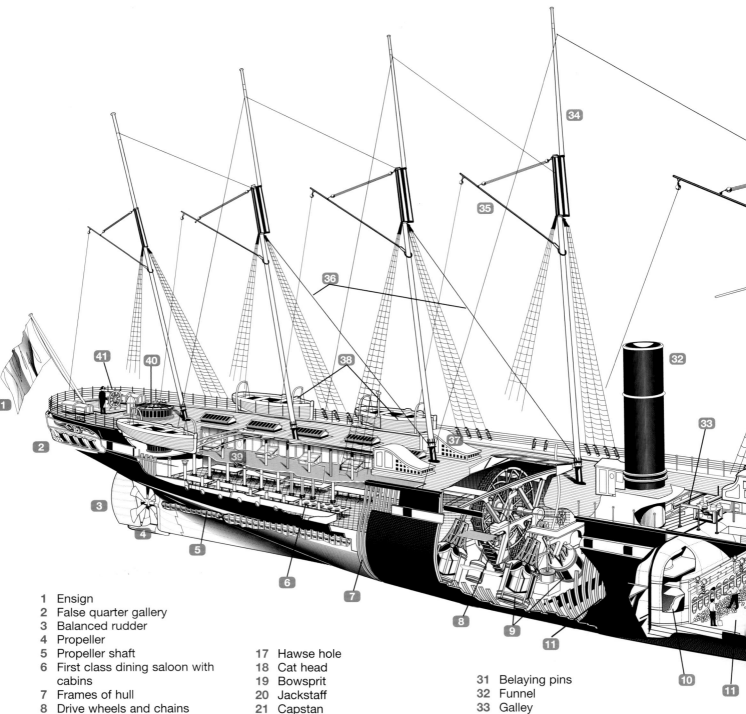

1 Ensign
2 False quarter gallery
3 Balanced rudder
4 Propeller
5 Propeller shaft
6 First class dining saloon with cabins
7 Frames of hull
8 Drive wheels and chains
9 Engine cylinders
10 Boiler
11 Boiler rooms
12 Steerage and second class bunks
13 Painted port on the side
14 Porthole
15 Crew's quarters
16 Anchor

17 Hawse hole
18 Cat head
19 Bowsprit
20 Jackstaff
21 Capstan
22 Forecastle deck
23 Forestay
24 Fore topmast stay
25 Shrouds
26 Deck lights
27 Windlass
28 Topsail yard
29 Main yard
30 Studding sail booms

31 Belaying pins
32 Funnel
33 Galley
34 Topmast
35 Gaff
36 Stays
37 Companionway to lower deck
38 Lifeboats
39 First class promenade saloon with cabins
40 Saloon light
41 Steering helm

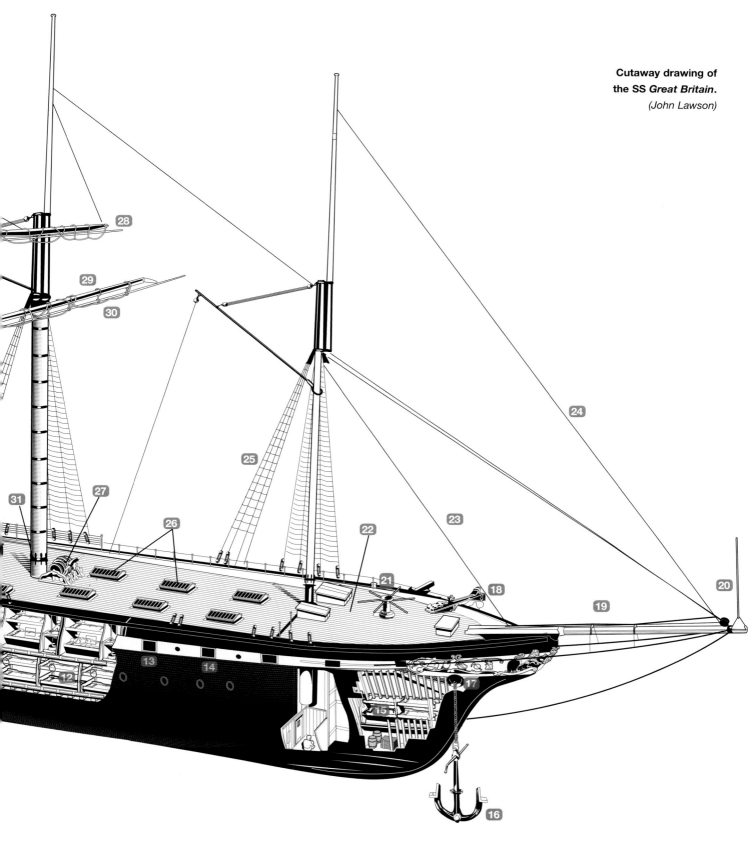

28

29

30

31

27

26

25

24

23

22

21

20

19

18

17

16

15

14

13

12

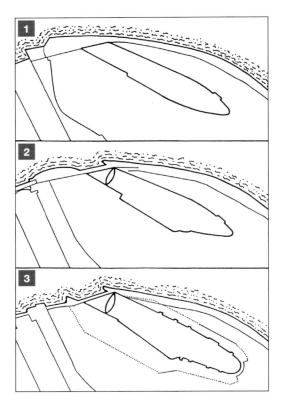

Building the ship

I t was not unusual to build large ships in dry docks, but it was rare to build a dock expressly for the ship. A site was chosen on the south side of Bristol's Floating Harbour, on largely undeveloped land, near a natural bend in what had once been the course of the River Avon, so that there would be space for the ship's great length when she eventually came out of the dock. It was leased from the shipbuilder George Hillhouse, whose Albion Yard was adjacent.

The dock was parallel to the line of the harbour, so it was near the waterside where materials could be unloaded. In its original form the dock closely resembled the shape of the *Great Britain*, but its head was later extended and rounded. There are many other inconsistencies in its shape. The western end, near the exit to the river, is in a curved section, close to where the caisson was fitted to seal off the water. Further along its length it has a

much flatter bottom, perhaps because it was deepened during the construction of the ship. The bottom rose during the building and Brunel advised Claxton, '. . . *of course* the bottom of the dock is likely to rise due to water pressure, they always do!' He recommended wooden supports and aged timbers can still be seen in the base of the dock. The shape of the south side was irregular, and became much more so during the 1850s when a further area was dug out, perhaps to build gunboats side by side.

The sides of the dock were fitted with steps for the workmen, and with horizontal ridges known as altars where the ends of the shores that supported the hull could be located. The ground between the dock and the harbour was not broken through until the ship was ready for launch. After that it was fitted with a caisson, a floating lock gate. The dock cost £53,000 and was ready late in 1839.

Indoor work was done in a large building near the dock, with each of its three storeys of diminishing height and round-topped windows

LEFT Timbers in the floor of the dock. *(Jonathan Falconer)*

LEFT Steps in the side of the dock. *(Jonathan Falconer)*

BELOW The caisson, which keeps the water of the Floating Harbour out of the dock. *(Jonathan Falconer)*

on the upper storey. The site was lit by gas and possibly had its own gasworks. A shipyard had a mould loft, a large, open room where the lines of the hull were drawn out full size and templates were made for frames and plates. This was usually on the upper floor, partly for light (which was less important with gas available) and partly because the lower floors were needed for heavy plant. The building was mostly destroyed by bombing during the Second World War.

The yard used advanced steam machinery bought from Naismith, Gaskell & Co. of Bridgwater. Possibly this was the equipment described by the French naval architect Dupuy de Lome in his report of 1842. There was a set of shears driven by hand or by steam for cutting the plates, of a design used through the 19th and into the 20th century. A 12ft combined piercing and shearing machine was less successful as it must have been very difficult to use, but a set of rollers for bending the plates was of a style that has never been superseded. However, it was limited in size to plates of

5ft 6in wide and ³⁄₈in thickness. There was a machine for bending the trough-like shape of the stem and stern posts, and hand ratchet drills for reaming out and drilling rivet holes. The machinery was valued at nearly £30,000.

Most of the heavy work was done in the open air. Covered building slips had been used in the Royal Dockyards, but there is no sign that the directors ever planned to use them as protection against the West Country weather. They paid for this during the winter of 1840–41, when the riveting 'was retarded during the frosty weather, when it was not prudent to proceed with it . . .'. There is no sign that any sophisticated cranes were used during construction; plates and frames were probably hauled into place with block and tackle.

The iron

The birth of iron shipbuilding depended very much on the development of the iron industry, and in particular on Henry Cort's system of 'puddling' in the 1780s. After smelting the ore, according to William Fairbaim,

in the furnace the iron is kept in a state of fusion, whilst the workman, called the 'puddler', by means of a rake or rabble, agitates the metal so as to expose, as far as he is able, the whole of the charge to the action of the oxygen passing over it from the fire. By this means the carbon is oxidised, and the metal is reduced to a tough, pasty condition, and subsequently to a granular form, somewhat resembling heaps of boiled rice with the grains greatly enlarged. . . . The puddler then collects the metallic granules or particles with his rabble, and rolls them together, backwards and forwards, over the hearth, into balls of convenient dimensions (about the size of 13-inch shells), when he removes them from the furnace to be subjected to the action of the hammer and the mechanical pressure necessary to give the iron homogeneity and fibre.

After that it could be rolled into plates, or angle iron. Abraham Darby of Coalbrookdale was commissioned to supply iron for the *Great Britain* (see box).

Iron for the SS *Great Britain*

The directors engaged the long-established firm of Abraham Darby of Coalbrookdale to produce the iron required to build the *Great Britain*. The company was enjoying a revival under the fourth generation since it was founded in 1709, and was rapidly adopting the latest technology. It had a hot blast furnace in its Horsehay site and had installed the latest type of guide rolling mill. In 1838 it exhibited two wrought iron plates, measuring 10ft 7in x 5ft 1in, in Liverpool, claiming they were the largest ever produced. The following year it produced 800 tons of plating for the *Great Britain*.

The plates were probably shipped out using a wharf on the River Severn that had been built in 1834. Iron was loaded on to vessels called trows, weighing up to 50 tons and carrying a crew of three or four.

Improvements to the Severn navigation were in hand, but locks and weirs were not yet completed and as late as 1846 '. . . the shoals and shallows . . . offer such serious obstruction to the important traffic of which it is the channel, that [there was a] number and variety of vessels so frequently to be seen stopped in their course, in many different places, for want of sufficient depth of water'.

The trows used currents, wind or horses to move downstream. After passing through the estuary they used the tide again to take them to Bristol Harbour, where they were a familiar sight. There does not seem to have been any great difficulty in the delivery of the iron, probably because the ship herself was delayed for many other reasons.

ABOVE Blast furnaces for the production of iron at Coalbrookdale, Shropshire, in the 1830s. *(TopFoto)*

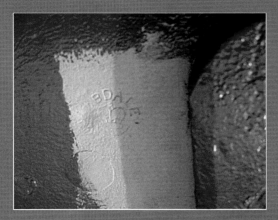

LEFT The Coalbrookdale stamp discovered on some of the ship's plates during restoration. *(ss GB Trust)*

The hull

Detailed accounts of the progress of the building of the ship are lacking, possibly because, as one man complained, 'The Directors take good care that the public should be kept in ignorance of their proceedings . . . they will not allow anyone to inspect the vessel or their works at Bristol without an order signed by two Directors and a contribution of 5s towards the sick fund of the workmen. Whoever heard of such a demand?' Annual reports to the shareholders make it clear that construction was well advanced by March 1840. A year later, progress was considered 'very gratifying' despite delays due to frost. The plating would be 'complete in two or three weeks' and was now being 'followed up with rapidity'. At the same time, 'a large proportion of the iron bulk-heads, beams, and smith's-work is already prepared, and will be fixed as soon as doing so will not interrupt the progress of the other works, and a considerable proportion of the deck planks is planked and placed in racks to season, and the joiner's work for the cabins has been commenced'.

Keel plates

Conventional ship construction started off with a keel which projected below the rest of the hull, and on which the structure rested during building. But like several ship designers of the time, Brunel was much influenced by bridge construction. Writing to Guppy, he emphasised 'longitudinal strength and stiffness' as essential features of the structure. The new ship had no keel in the conventional sense and used a more complex system than was normal for the time. The basis of the ship was a girder structure as described by Brunel's son:

Over the traverse iron ribs at the bottom of the ship were laid ten deep longitudinal beams . . . which, over the greater part of the bottom of the ship, were covered with an iron deck riveted to their upper edges by angle irons, thus forming a cellular structure which added greatly to the strength of the ship. It does not appear that this deck was designed to be water-tight, so that it did not form the same security against accident as the inner skin of the cellular structure which Mr. Brunel later adopted in the 'Great Eastern'.

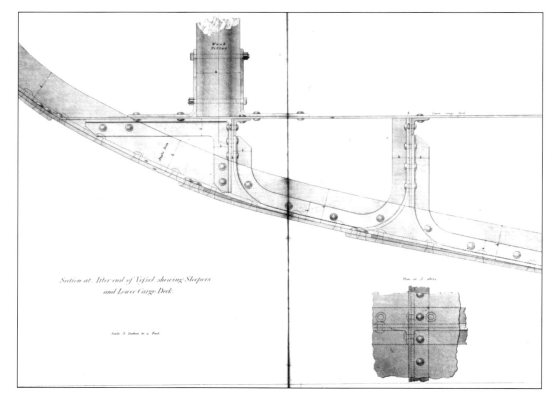

RIGHT The structure of the bottom at different points: after end of the ship showing sleepers and cargo deck.
(ss GB Trust)

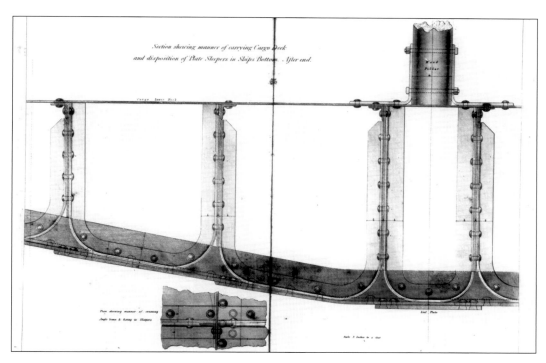

LEFT **After end showing the manner of carrying the cargo deck and disposition of plate sleepers in the ship's bottom.** *(ss GB Trust)*

The keel plates were ⁷⁄₈in thick and 20in wide, welded into lengths of 50–60ft, joined together with scarfs or overlaps of 1ft 6in. The thickness was increased to an inch towards the bow and stern as being 'more liable to touch the ground'. Above them were ten U-shaped girders which formed the main strength. These were 3ft 3in deep and ½–⁷⁄₁₆in thick. Those in the centre ran the full length of the bottom, others terminated where the hull rose at the turn of the bilge. They were backed up by strong angle irons running across the hull.

FAR LEFT AND LEFT **The bottom structure as it survives today.** *(Jonathan Falconer)*

FAR LEFT **Curved frames inside the forward hold.** *(Jonathan Falconer)*

LEFT **The tank top.** *(Jonathan Falconer)*

Frame and skin

In the most fundamental sense, the *Great Britain* followed the principles of western ship design – she had an inner frame of curved ribs (known as timbers in the case of a wooden ship) and a skin (made up of planks in a wooden ship, plates in a metal one). The use of iron meant that the relationship between the frame and the skin was altered. Unlike planks, each plate was joined to the one next to it to create a far more rigid structure, and one which was less dependent on the frame to give it shape. The *Great Britain* used the system of 'clinker-building', which went back to the Anglo-Saxons. With wood there was a limit to the size of such a vessel. King Henry V had used a triple-skin system for his 1,500-ton *Grace Dieu* in 1418, but that was taking the process beyond its limits. Later wooden ships were built in the carvel system, in which the planks were laid side by side. But iron plates could be much thinner than wooden planks, and it was already

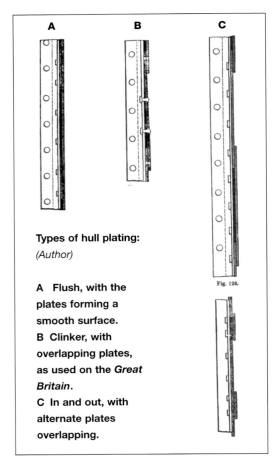

Types of hull plating:
(Author)

A Flush, with the plates forming a smooth surface.
B Clinker, with overlapping plates, as used on the *Great Britain*.
C In and out, with alternate plates overlapping.

normal to rivet them together. Alternatives included laying the rows, or strakes, of plates side by side as in wooden construction, then covering the gap with an internal strake of narrower plates. It had the advantage that the surface was flush in the final result, but unless the inner strakes were double-riveted to both plates, the result was weaker than clinker. According to Claxton:

The Great Britain is built with lapped joints in preference to flush, the first system adopted in iron ship-building, representing carvel-built ships. The lapped joint is the method employed for clinch or clinker-built vessels. Trials were made at the Company's Works of the comparative strength of the two methods, and the lapped joint was stronger by one-fifth of the whole strength. It is obvious for the purpose of resisting lateral pressure or the blows of the sea, on the broadside, it must be better than the flush system, where all the strain must be thrown upon the ribs, beams and decks, the latter horizontally; while, with the over-lapping joint, in addition to that resistance, the plates themselves bear against each other and assist each other in resisting a pressure great in proportion to the length of the vessel. . . .

In flush jointing, the butt plates inside would, for double riveting, have to be twice the depth of the lap of the joint in the other system, consequently a great additional quantity of iron would be required for each seam, or in the Great Britain about 18,000ft of iron, 6in by ¾, and double the number of rivets, in addition weight of nearly 100 tons.

As in all iron ships the butts, or vertical joins between the plates, were staggered so that there was no single line of weakness in the length of the hull.

One of the great difficulties was the small size of iron plates that could be produced economically – no more than 6ft 6in long and 3ft wide. Since they were held together by double rows of rivets, the size was reduced to about 6ft. This was not a multiple of the distance between the frames, which was supposed to be 18in in midships increasing to 24in towards the bow and stern – though this was highly variable, because the frames were carefully arranged to avoid the vertical joints, or butts, between the plates. This raises the question of whether the *Great Britain* was built 'skin first' like a Viking ship, or 'frame first' like nearly all western seagoing ships for at least 500 years. It might have been possible to build the ship skin first, using a cradle outside or inside the hull to support the structure until the frames could be fitted. But this would have caused difficulties in the lower part of the hull, where holes had to be cut in the longitudinal structure for the frames. Moreover, such an unconventional method would probably have excited some comment.

It seems more likely that the building was planned carefully in advance, using plans or models to find the exact position of each butt, and then placing and shaping the frames to fit in with that. John Grantham wrote later: 'The builder now prepares what he calls a block model . . . upon the surface of which he draws every plate and frame, being regulated in this by

LEFT The system of riveting as used on the *Great Britain*'s hull. *(Jonathan Falconer)*

BELOW LEFT The interior of the hull showing the rather irregular arrangement of frames and butts. *(Jonathan Falconer)*

BELOW The framing of the hull from inside, showing the structure of the bottom with a stringer running fore and aft, and pillars to support the deck above. *(Jonathan Falconer)*

the specification which has been given him. . . . A mark is then put to every frame and plate, and a list made of the size and thickness of each.'

Often the riveter had to drive in a tapered steel drift to modify the shape of the holes. 'He then inserts the rivet, which is hammered up in the usual way. If the rivet is very hot, and the hole not very irregular, by dint of hard blows the hole is well filled, and the work sound, though about this there is no certainty.' But Grantham had nothing but praise for the work on the *Great Britain*.

The lower plates of the ship were fitted in 'reverse clinker', with those nearest the centre line overlapping those further out. This implies that the first strake to be fitted was the one at the turn of the bilges, which was then overlapped on both sides. In many ways it would have been the most difficult one, curved throughout its length. It would not have been easy to fit the plates further in, as the workers would have had to get at them from underneath and support them in place while they were riveted; but the ship was high enough in the dry dock to do that. There were four courses of plates $^{11}/_{14}$in thick, that is those fitted by reverse clinker in the lowest part of the hull. The plates above that were thinner, with several courses of $^{5}/_{8}$in, reducing to $^{3}/_{8}$in at the highest level of the hull.

In riveting, ' . . . the usual practice is to punch all holes in the under side of plates to a templet, observing always that the side from which the rivet is inserted is the lower side when under the press. This causes a slight

taper, which punched holes always have, to be in the right direction for the insertion of the rivet.' After one strake had been fitted, the holes for the next one were located.

When these plates have been fitted, those for the outer strake are applied, and the exact position of each hole is marked by inserting through the inner plate a round plug, previously dipped in some liquid whiting, or an elaborate and rather complicated templet is made from these holes, by which the outer plate is punched. The plate being, in either process, thus prepared for punching, is taken to the press, and being lifted and guided by a gang of men, each hole is separately punched. Any one who has watched this process with heavy plates knows how impossible it is to form the hole exactly where marked; and I may safely say that however careful the men may be, the holes, as a rule, are never fair with each other.

The frames were bent from L-section or angle iron, 6in x 3½in and $^{5}/_{8}$in thick. The rivets holding the plates in position were fixed to the smaller, 3½in edge. Clinker construction would naturally leave a narrow triangular gap between the plate and the frame, so pieces of iron known as liners were carefully shaped and inserted to fill it. The plates had to be cut to shape very accurately, especially in the difficult areas near the bows and stern, and this was the job of the shearing machine. The relatively flat ones could probably be bent into shape *in situ*, but many would have to be shaped in the rolling machine – that was a craft which required a good deal of skill, especially when the plate was to be curved in more than one direction. The structure was stronger in the engine and boiler room areas, where the ribs were doubled with more angle irons riveted to them on the inside. The lower ends of the pistons rested on bed plates in the turn of the bilge.

Outside the hull, the ship was fitted with two 110ft bilge keels near the turn of the bilge. They were intended to steady the ship during docking, but they also had a role, unrecognised at the time, in reducing rolling. Each was made up of a 1¼in thick plate, supported by an angle iron on each side.

The visiting French engineer Dupuy de Lome was not impressed with the structure, according to his report of 1842, possibly because he did not understand the role of the bottom and deck beams in strengthening it.

. . . I found the fabric too weak in many parts, in fact the floor timbers are not strong enough for the size. It is only two angle irons one against the other, and the joinings are thus doubled only as far as the first deck, of thin iron plate, which is 3ft 3in above the keel, and from that to the gunwale is a single angle iron, except in the part about the engines, where they are doubled almost to the top. It is true that this angle iron is 6in by 3½in but that is nothing, this plate could not resist a shock even going at a moderate speed, the [ribs] are of single angle iron of 9in by 3in – it is very weak in proportion to the size of the vessel.

In fact the structure was very well built. Operations such as rolling plates and the fitting of liners were often bodged in ships of the period, but those of the *Great Britain* were carried out with great care. Unlike a wooden

ship she did not need to be caulked by driving oakum between the planks, and she remained dry throughout a long and hard life.

Bow and stern

The foremost part of the bow was based on a trough-shaped piece of iron known as the stem, 12in deep at the forefoot where it joined the keel structure, and 5in thick. It declined gradually to 1ft deep and 1½in thick. It was welded in a single piece and had a double curve, which formed the distinctive shape of the bow.

ABOVE The bilge keel underneath the hull. *(Jonathan Falconer)*

LEFT The upper cargo deck (which is missing) would be supported by iron brackets. *(Jonathan Falconer)*

The reports are generally vague about the more complex structure of the stern, and it was heavily modified to accommodate the lifting propeller in the 1850s. The lowest part was something like a normal keel and it extended well aft of the propeller to hold the foot of the rudder. Below the waterline, the structure was based on a vertical iron post 14in fore and aft and 2in wide, with a boss to take the propeller shaft. It is wider at the top, a relic of the days when the ship had been intended for paddle wheels, and before an aperture for the propeller was needed.

One of the most innovative features was the balanced rudder. Ever since the 14th century, western ships had been fitted with rudders that hinged on their forward edge, against the stern post. The placing of the screw perhaps allowed some rethinking of this, and Brunel designed a rudder that was pivoted about a third of the way along its width. This meant that it would be far easier to turn, but its snag, in the eyes of the seamen of the day, was that it would not automatically go into the neutral position if there was not pressure on the helm

Above that, the shape is rather traditional,

ABOVE The bow showing the hawse pipes through which the anchor chain was let out and taken in. There are two sets of hawse holes – the lower pair are the original 1843 hawse holes, which were plated over and replaced with the upper pair in 1852, when the ship sat lower in the water after her major refit. *(Jonathan Falconer)*

RIGHT Stern with propeller and rudder. *(Mandy Reynolds)*

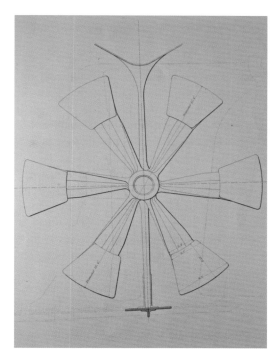

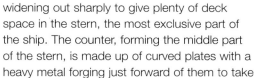

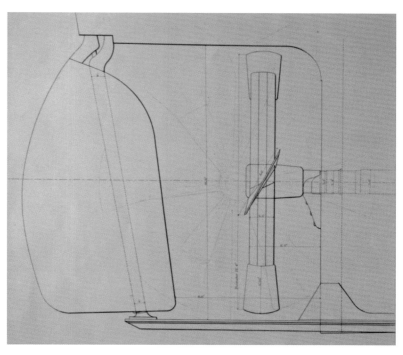

widening out sharply to give plenty of deck space in the stern, the most exclusive part of the ship. The counter, forming the middle part of the stern, is made up of curved plates with a heavy metal forging just forward of them to take

LEFT AND ABOVE The six-bladed propeller and the rudder behind it, showing how it rotated aft of its leading edge to provide balance and make it easier to operate. *(ss GB Trust)*

BELOW Stern and quarter. *(Jonathan Falconer)*

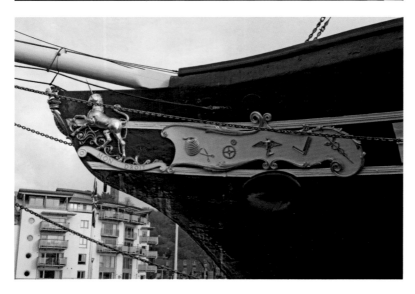

the head of the rudder. Above that the surface is flatter and more vertical, and fitted with ten windows for the saloon passengers' comfort. There are false projecting galleries on either side to give a more traditional appearance, with the frames of imitation windows on them.

The decoration of the ship was 'like most of the Bristol ships, plain but substantial in finish', according to Claxton. The 'figurehead' was a royal coat of arms, backed up by a lion and unicorn. Aft of that were the trailboards with symbolic decorations representing different professions – a gear wheel for engineering, rod of Asclepius for medicine and so on. Aft, the moulding of the false galleries ended in cornucopia, a common motif on ships. The flat of the stern was surmounted by a version of the Bristol coat of arms, showing a ship passing a castle, supported by unicorns.

Decks and bulkheads

Each of the ship's five decks was different in construction. The top of the double bottom, known as the lower cargo deck, was an integral part of the lower structure. At the next level, the upper cargo deck was supported by angle iron beams of 6in x 3in and ½in thick, like all the others. Unlike them it was plated with $\frac{5}{16}$in iron, with $\frac{9}{16}$in thickness round the sides. Like all the decks the beams curved down slightly to meet the ribs, and it had diagonal brackets in the angle with the ship's side, though these were relatively weak. The deck above, the lower deck, was made up of 5in pine timber, laid across the hull instead of along it in more normal fashion. It had to bear a good

OPPOSITE a section of the hull was reconstructed by Ewan Corlett. *(ss GB Trust)*

STEAMSHIP GREAT BRITAIN

GENERAL STRUCTURAL ARRANGEMENT 1:24
AS BUILT 1843

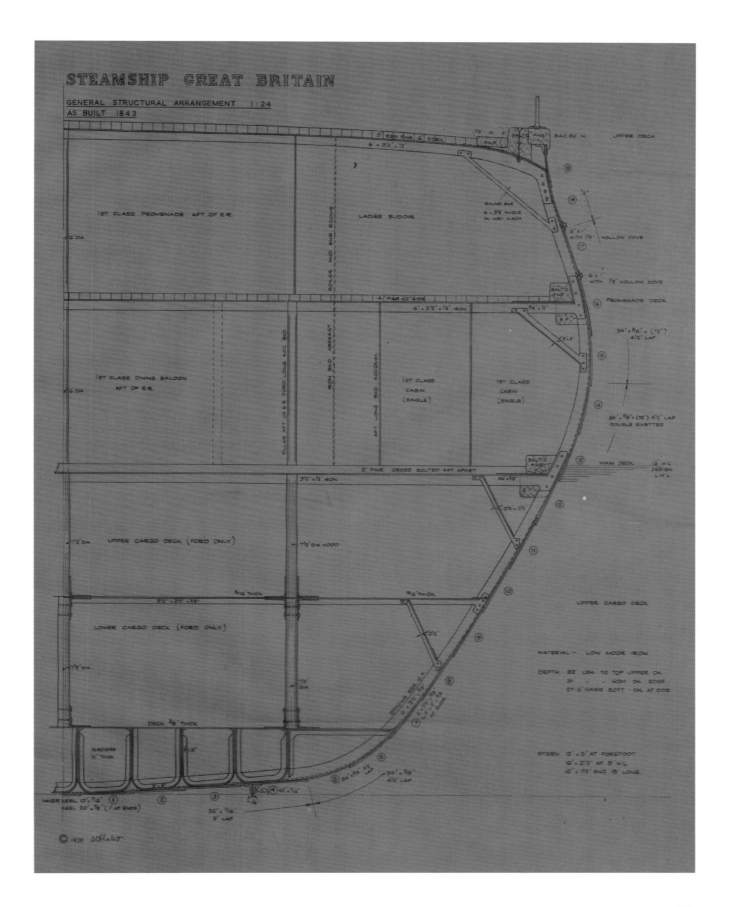

deal of stress in preventing the sides being pressed in by the force of water, and according to Guppy,

> as the deck is situated on the load floatation plane of the vessel, where transverse stiffness is of more importance than longitudinal strength, the planks are placed athwartships, and their extremities are firmly bolted down, through two longitudinal stringers of Baltic timber, to the shelf plates, which are three feet wide by $^3/_8$in thick, and are very securely fixed to the sides.

These stringers were raised above the level of the deck to help prevent any water from reaching the sides. It might have added strength to make this deck of iron like the one below, but it was used more by passengers and crew and as a visitor joked feebly in the *Mechanics Magazine*, 'To decks made of iron plates I think the sailor would have a very *standing* objection.'

The middle or promenade deck was laid with 4in pine, in more common fore-and-aft fashion, as it bore less stress. Above that was the upper deck, which was the only one with any camber, or curve. It was unusually important in that it supported the whole of the masts, as they did not pass through it to other decks as in conventional ships. Each beam was constructed almost like an arch, curving down towards the sides of the ship, though with a sharp bend where it met the ribs. Guppy treated it as a problem in civil engineering, in that '. . . the sides of the vessel at this height, and also this deck, may be considered as the truss, which is to resist longitudinal deflection, or drooping of the extremities . . .'.

Despite the obvious advantages, wooden ships had not been fitted with bulkheads to divide them into watertight compartments. In March 1843 the shareholders were told: 'For the sake of adding greatly to her strength, and diminishing the apprehension of fire on the part of passengers, the decks and partitions of the body of the ship, occupied by the engines and boilers, will be fitted up in iron instead of wood work. . . .' There were five bulkheads, made of iron plates staggered like brickwork. The first was like a modern collision bulkhead. '. . . As it is in the forepart of a vessel that injury is most likely to be sustained, this partition is made particularly strong and secure.' Two others were situated fore and aft of the engine and boiler rooms. The remaining two were in the spaces between these and the bow and stern, but they were only carried up to the level of the lower deck. Dupuy de Lome was not satisfied and wrote to the Merseyside shipbuilder John Laird:

What I would not above all have in a vessel built for us, is that the traverse partitions destined to divide the vessel into independent parts, only reach as far as the deck, at 3 feet above the keel, this bridge (or deck) ought to be joined to the sides of the vessel in such a manner as not to let the water pass. . . .

There were also fore-and-aft bulkheads in the area of the engine and boiler rooms, curved inwards between lower-deck and middle-deck level, creating extra strength and helping to seal off the coal bunkers. The decks were supported along the centre line by a series of wooden pillars, 7½in in diameter.

John Grantham wrote in 1842: 'The general character of the workmanship is very good, and does great credit to her builder.' Dupuy de Lome, on the other hand, wrote to Laird: '. . . it is very inferior to what I have seen of yours – as much in <u>plan</u> as in <u>finish</u>.' The local

civil engineer J.R. Hill agreed that 'The whole of the materials and workmanship, both of ship and machinery, appear to be of the first order.' The survival of the hull for more than a century and a half suggests that Grantham and Hill were closer to the truth.

On deck

Arriving on the upper or weather deck, a visitor of the 1840s would immediately be impressed by the amount of space. It was very long, and the tumblehome amidships combined with the flare of the bows to make its sides almost parallel. Unlike earlier ships, it did not

ABOVE Iron pillars supporting the deck from underneath, in the fore end of the ship. *(Jonathan Falconer)*

BELOW The upper deck looking forward showing the deck lights, the windlass, and a modern ramp for visitor access. *(Jonathan Falconer)*

ABOVE The forecastle deck, showing a companionway leading to the crew's quarters below, the cathead, the stout bitts used for securing ropes and the anchor cable, and the ship's bell behind them. (Jonathan Falconer)

RIGHT Wheel and binnacle. (Jonathan Falconer)

BELOW The circular light above the main saloon. (Jonathan Falconer)

BELOW RIGHT Engine room deck light. (Jonathan Falconer)

have a raised quarterdeck in the stern and the forecastle in the bows was only about 3ft above the main deck. The deck was not interrupted by rows of guns like the East Indiamen, which had carried out long-distance passenger travel until recently.

Unlike later ships it had no superstructure in the sense of erections that were high enough to create cabins and other usable spaces underneath them, though it did have many

projections to interrupt the space. These included the ensign staff projecting over the stern. Just forward of that was the steering wheel, located in front of a small deckhouse which contained its gearing. There were small deckhouses on either side, containing flags, navigational equipment such as log lines and spare rigging equipment. Immediately forward of the wheel was the compass in its binnacle, surmounted by a curved head. Forward of that was a large round skylight with four smaller, rectangular skylights on each side of the two aftermost masts, to give the saloon natural light. There was a capstan between them.

Opposite the next mast, the fourth, were two curved companionways covering the steps to the decks below. Forward of that mast was the largest structure on the deck, the skylight above the big wheel of the engine gearing. Then came another small skylight, the funnel (which was often a meeting place for passengers) with chicken coops and a cowshed nearby. There were two more companionways, then the forward part of the deck had four more small skylights on each side for the forward saloon. A windlass was sited just forward of the main mast and the foremast was just aft of a small raised forecastle, which contained another capstan and gear for handling the anchor.

If any of the passengers were disconcerted at the idea of sailing in an iron ship, they would not have been reassured to notice that the lifeboats hanging from the sides were also made of iron. As described in 1847 they consisted of 'Three Iron LIFEBOATS, one capable of carrying largely and floating

lightly, being a flat. One, a fast sailing boat of beautiful form, but little the worse for wear.' Mr Dearlove, a passenger in 1862, was not satisfied and commented, '. . . the very contemplation of a Fire on board this vessel is awfull as the boats could not carry above half the people aboard, although their is eight boats, that is not sufficient to contain 800 souls, therefore about half would have to be left to their fate.'

Below decks, the ship was divided in two halves by the engine room and boilers, and also by function and social class. Going down one of the companionways aft of the engine room, the visitor would enter the promenade saloon. There were cabins on each side, arranged in pairs with one at the ship's side and one inboard, both accessible by a small corridor, as described in the *Bristol Gazette and Advertiser*:

> *In this part of the ship there are twenty eight state bedrooms, all single, and these possess an advantage over those in any other vessel, inasmuch as ladies pass from their own private saloon to their bedrooms*

ABOVE LEFT **The funnel, supported by wire stays.**
(Jonathan Falconer)

ABOVE **Companion-way next to the engine room leading aft.**
(Jonathan Falconer)

BELOW **An 1845 view of the promenade saloon.**
(Illustrated London News)

RIGHT Deck light seen from inside the promenade saloon. *(Jonathan Falconer)*

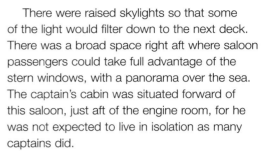

BELOW Stern windows in the promenade saloon. *(Jonathan Falconer)*

BOTTOM The promenade saloon, with passenger cabins leading off on either side. *(Jonathan Falconer)*

without passing through the promenade, or being seen by those who are there. . . . The door of each bedroom instead of heretofore opening direct into the promenade or saloon, now opens into a small passage so that each room can be approached and left without the interior being visible to the parties outside.

There were raised skylights so that some of the light would filter down to the next deck. There was a broad space right aft where saloon passengers could take full advantage of the stern windows, with a panorama over the sea. The captain's cabin was situated forward of this saloon, just aft of the engine room, for he was not expected to live in isolation as many captains did.

The next deck down was one of the showpieces of the ship, the dining saloon. It too had cabins on each side, but it was at the maximum breadth of the ship and was wider and more regular than the saloon above. The correspondent of the *Bristol Gazette and Advertiser* was filled with local pride during a preview, though he seems to have confused this saloon with the one above.

. . . the deck above is supported by twelve principal columns in the centre with fancifully carved capitals and twelve smaller ones ranged on each side, while the side will be decorated with ornamental pilasters, the whole in arabesque style. The door to each state cabin is carved and gilded, and it is also surmounted with carved work, the whole of the fittings present a most elegant and luxurious appearance.

Even Queen Victoria was impressed: 'The Dining Saloon is immense & the lounges for passengers, on both decks, are very large.' But the correspondent of the *Brooklyn Eagle* had seen the latest American steamships:

Upon the whole, we do not think the arrangements for comfort exceed those on our North River boats – the staterooms are certainly no larger or more commodious; and, as there is no accounting for taste, we would prefer a sojourn upon the Columbia or Niagara to one on board the Great Britain.

The *Illustrated London News* claimed that there was 'little in the way or ornament' in any of the saloons and that the style was of 'plainness and simplicity', deploying 'rigid utilitarianism' in which 'not one inch is thrown away'.

Under this, in the 'cauldron below' according to the *Illustrated London News*, were two decks for cargo and coal. The engines and boilers took up almost all the central part of the ship. Below decks, there was only room for passages on each side on the upper deck, with bathrooms and toilets on one side and the galley or kitchen on the other. 'This is most completely fitted with every conceivable apparatus for roasting, boiling, frying, grilling and stewing; sending forth to all who can on a sea voyage enjoy them, a series of excellent and tempting dinners.'

The passenger accommodation was similar forward of the engine room, according to *Chamber's Edinburgh Journal*. 'Except in the shape, we saw nothing in the plan of this different from that of the after-saloon. It is of course much shorter, by reason of the sudden narrowing of the ship where the prow is formed.' Below that were more cargo decks and coal storage. Forward of the bulkhead in the bows, and thus separated from the passengers, were the crews' quarters. The officers, apart from the captain, lived at the top level, under the forecastle deck, with the seamen and stokers below that, in a compartment of triangular shape because of the narrowing of the bows. They slept in bunks rather than the traditional hammocks.

During her major refit of 1852 the ship acquired a deckhouse in the stern, which was luxurious enough to satisfy even the *Illustrated London News*. It was

> . . . tastefully decorated with paintings on glass of the armorial shields of all the nations. Two ranges of tables run the whole length. Opposite the door is the pantry and at the further end is a music room with an ornamental skylight in which the deck binnacle is placed. Behind is the smoking room and still further aft the wheelhouse.

During the next refit in 1857 this was developed into a full deck, extending from one side to the other and along the full length of the ship. The layout of the deckhouse remained similar and first-class passengers now lived aft on that deck and with a lower saloon below, with second-class passengers just forward of them. More second-class passengers were accommodated forward of the engines, on

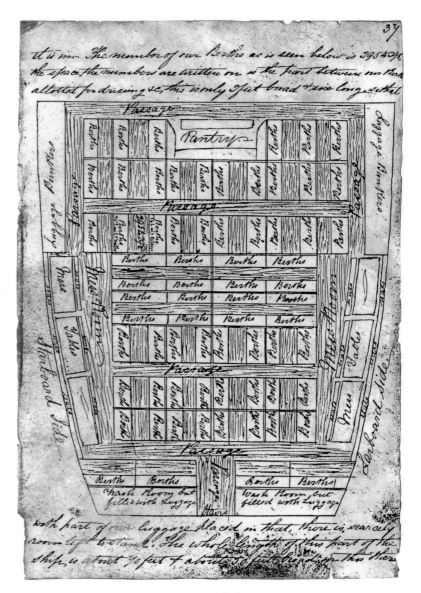

both decks. This left the former cargo decks, fore and aft, for either cargo or emigrant or 'steerage' passengers, in simple wooden bunks that could be erected or dismantled as required. Conditions were far from luxurious. According to Allan Gilmour in 1852 steerage was in 'a crowded condition' and each of the numbered berths was 'only two feet broad and six long, so that with part of our luggage placed in that, there is scarcely room left to stand. The whole length of that part of the ship is 70 feet and about 50 feet broad, in this there are crammed 104 berths. The only sitting room we have is the Mess Room, where also we get our meals. This room is not large enough to accommodate more than half the passengers. . . .'

ABOVE A plan of the berths in steerage during the Australia run, drawn by passenger Allan Gilmour in his diary. *(ss GB Trust)*

Wire takes the strain

An innovative feature of the *Great Britain* was the use of wire rigging, which seemed unbelievably light to contemporaries who were used to much thicker rope as can be seen on HMS *Victory* today. The standing rigging was fixed in position to support the masts, in contrast to the running rigging which was used to adjust the sails. It consisted of the shrouds that ran from the outside of the hull to near the head of the lower masts, and braced them against forces from behind and the sides, the backstays aft of them, and the stays which ran diagonally forward of the masts.

The *Great Britain* has five shrouds on each side of the second mast and three on each of the others. Each shroud had the traditional system of deadeyes and ropes known as lanyards to tighten it, though wire is far less likely to stretch than rope.

Each set of shrouds had horizontal ropes known as ratlines to allow the crew to climb into the rigging. Behind these were adjustable backstays, which did a similar job to the shrouds but were set further aft – they would have to be let out on the leeward side to avoid interfering with the sails, while their partners on the other side were tightened to support the masts. The forestays were in pairs running from the head of each lower mast to the deck just aft of the next mast forward.

Wire rigging was largely untested at the time and treated with suspicion by seamen. To Brunel, its greatest advantage was that it created much less air resistance when sailing or steaming into the wind.

BELOW The shrouds and lanyards, which support the masts from the side and rear. *(Jonathan Falconer)*

RIGHT Original deadeyes that were salvaged from the ship. *(Jonathan Falconer)*

BELOW Replica deadeyes, showing how the ropes known as lanyards were threaded through them and used to tighten the shrouds *(Jonathan Falconer)*

BELOW RIGHT A view from below showing the ratlines, which formed steps for the crew to climb the masts. *(Richard Pope)*

Rigging

The *Great Britain*'s six masts made her unmistakable from a distance, even before the hull of the ship had appeared over the horizon. Once on board, a visitor saw more unusual features. Each mast was fitted in two sections in more or less traditional style. The upper part of each, the topmast, was set just forward of the lower mast and overlapping with it; the two were joined together by caps at the level of the bottom of the topmast and the head of the lower mast. They did not have the large platforms, known as tops, of a more conventional rig, which helped to spread the shrouds of the topmast – for the topmasts were usually lowered if not in use.

New terminology was needed for the *Great Britain*. Formally there was the fore, main, mizzen, jigger, driver and spanker. Less formally the seamen called them Monday, Tuesday, Wednesday, Thursday, Friday and Saturday – conveniently the name of the Sabbath was not needed in such a God-fearing age.

The original masts were hinged just above the deck level, which was common on the barges that Brunel must have known when he worked on the Thames Tunnel. They obviously found this essential when 'shooting' the numerous bridges of the river, but that was not useful in an ocean-going ship. Because of the propeller shaft the after masts could not be stepped on the keel in conventional fashion, so that might have been a factor in the decision. Brunel may have considered them useful in reducing top-weight in a storm – the *Illustrated London News*

suggested that they might be 'struck in a gale of wind with the greatest ease', but most seamen preferred to retain some capacity to set sail and control the movement of the ship.

Claxton claimed the hinged masts were intended to reduce resistance when steaming into the wind, '. . . when, in the Captain's judgement, contrary gales shall appear to have set in, as the westerlies do at certain seasons of the year. . . .' But not all sailors would have agreed, and even Claxton was defensive. 'To a seaman's eye they have a look of insecurity; but if the strain which a fixed mast will stand is compensated by additional shrouds and stays, either in strength or quantity, the same end is attained.'

Not everyone was impressed with the overall appearance, particularly in New York.

> . . . her great length was apparent, but I did not think she looked very elegant, with her six low masts. The graceful rig of the American vessels is well known, and the *Great Britain* contrasted unfavourably in the opinion of many.

The changing rig of the *Great Britain* reflects her increasing use of sail rather than steam. As early as 1846 her third mast was removed so that she was fore-and-aft rigged on numbers one, four and five and square rigged on two and three, with rather stronger and better supported topmasts. In 1852 mast number 3, aft of the funnel, was removed so that she was fore-and-aft rigged on masts one and four (according to the new numbering) and square rigged on two and three. A year later, when it was clear that she should follow the sailing ship route to Australia using steam only as an auxiliary, she was rigged like a clipper ship, with three masts, all square rigged.

Engines

The correspondent of *Chamber's Edinburgh Journal* provides a view of the power plant in June 1845, although his terminology is confusing to the modern mind because 'engine' often meant the same as 'cylinder'.

The bewildering mass of machinery which meets the view – bewildering rather from its vastness than from its complexity – is so distributed in the vessel as to form permanent ballast; and the middle of the ship has not to bear all its weight, as in the case in vessels with side paddles. . . . Four engines – the united power of whom equals that of 1000 horses – are employed. . . . There is an enormous wheel or 'drum,' 18ft in diameter, working on an axis or spindle. To either end of the spindle is attached an immense crank moved by one pair of engines, the other pair driving the crank at the opposite end of the axis; so that the whole four expend their force upon the gigantic drum to whirl it round. The duty of this large wheel is

to cause a band composed of four iron chains to revolve with it; that in the regions below, they may pass round and turn another and smaller wheel. . . . Though weighing seven tons, therefore, the chains work quite silently.

Queen Victoria found the engines 'beautiful' in April 1845. Passengers occasionally visited the engine room by invitation, and one described a visit in 1862. Though they were not the same as the original Brunel engines, the effect on a visitor was probably similar.

Entering at the door on the main deck, we descend by an iron ladder to the first flat, a gallery around which are situated the engineers' rooms. Descending another flight of iron steps, we find ourselves in the immediate presence of the engines, to which our attention is now directed, the eye being attracted by the two powerful pistons as they rise and fall in their mysterious sockets. There are two complete engines (both of the same dimensions, and alike in every respect) employed in driving the screw.

BELOW Details of the *Great Britain*'s first engine. *(ss GB Trust)*

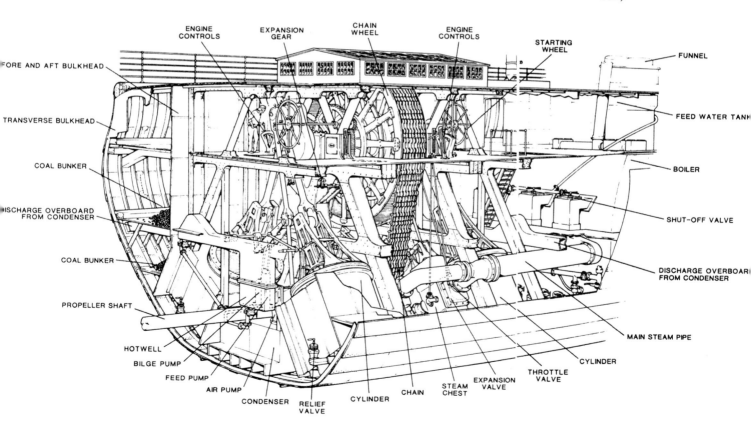

K. A. GRIFFIN

Chapter Three

Sailing the *Great Britain*

The captain, or master, of a ship was expected to be a seaman, a navigator, a disciplinarian, a diplomat and a leader. He had the power to make life on board pleasant and profitable, or introduce discontent and tyranny. Fortunately, none of the SS *Great Britain*'s captains fell into the latter category, although some had their faults.

OPPOSITE The SS *Great Britain* under sail and steam at sea in **1845.** *(Keith A. Griffin/ss GB Trust)*

Captains

The captain of a merchant ship was technically known as the master, the title of captain was a courtesy one. He was all-powerful, especially on an ocean-going vessel that might spend many weeks without communicating with the outside world. According to the nautical writer Richard Dana, 'As soon as the pilot leaves the ship, the entire control and responsibility are thrown on the master.' He was expected to be all-seeing and all-knowing – a seaman, a navigator, a disciplinarian, a diplomat and a leader.

'He has a power and an influence, both direct and indirect, which may be the means of much good or much evil. . . . He may make his ship almost anything he chooses, and render the lives of his officers and men pleasant and profitable to them, or may introduce disagreement, discontent, tyranny, resistance. . . .' Fortunately none of *Great Britain*'s captains fell into that category, though some had their faults.

Her first captain, James Hosken, was born in 1798 and went to sea at the age of five or six with his father, a naval warrant officer. He 'used to be perched on the table to go through the forms for ordering the ship to be put about &c.'. He joined the Royal Navy more formally as a midshipman in 1810 and came tantalisingly close to action, but the war ended in 1815 before he had completed the six years needed to become a lieutenant. He had to serve as 'passed midshipman' until 1828, when he was finally commissioned. In 1832 he joined the packet service carrying overseas mail, then took command of merchant ships. He captained the *Great Western* on 32 voyages, including her historic maiden transatlantic voyage in which she raced the *Sirius* to make the first steam crossing of the Atlantic. Speed had perhaps become an obsession with him – he put the *Great Britain* in unnecessary danger more than once. He was popular with the crew and passengers, but his navigation left much to be desired.

Barnard Mathews commanded the ship during her first Australian voyages. A Cornishman, born in 1803, he had already captained the *Great Western* during several voyages. Anne Henning found him 'a most merry, good tempered man who always keeps his end of the table in a roar'. But Seaman John McFall lost confidence in him during the first Australian voyage: '. . . If they had a competent commander we should have been well on between the Cape and Australia. This one, I am sure, is not fit to be here.' He was offered the lucrative post of Lloyd's agent in Melbourne and travelled out with his family, acting as supernumerary captain to guide the new appointee.

John Gray was born in Shetland where he learned to steer the local fishing craft at an early age. He was already in command of a sailing ship when he was asked to join the *Great Britain* as second mate in 1852. He was not happy at first, as he could not get through his work and he did not like looking after the second-class passengers. Nevertheless, he was promoted to first mate and Anne Henning found him 'a very nice man, married of course'. Gray proved to be an ideal captain when he was appointed in 1854. He was a good seaman, a firm but fair disciplinarian and very attentive to the passengers. To the cricketer William Caffyn he was 'a remarkably nice man, and endeared himself to all of us by his many little acts of kindness'.

Every captain is fond of his ship unless something is seriously wrong, but Gray went further when he told Mary Compton, 'I love every plank of her. I pat her sometimes. . . .' He was a man of considerable discretion. When a Mr Shanassey brought 'some rough Irishmen' to a first-class table to play cards,

> . . . the captain was very indignant about it
> but did not want to speak himself because it
> would have created so much ill feeling, but
> he told Miss Farrar and me that directly tea
> was over to take the seats at the head of
> the table and not to stir for anybody, we did
> so and got Mrs. Fenwick to join us and we
> spent a very merry evening playing at games.

Gray lived in Liverpool with his wife and children. When some of the family visited the ship on her return in 1870, John D. Parr remarked on 'The Captain's wife and three lovely girls' who had a 'loving scene' on the

deck. Yet in 1873 he disappeared from the ship during a voyage home in an apparent suicide caused by illness and depression after the death of his only son.

Charles Chapman was appointed captain. He had been born in Lincolnshire in 1821 and gained his master's certificate in 1854. He had served six years as mate on the *Great Britain* before becoming captain of a fast merchant ship, the *Thetis*, but he returned to the great ship as mate in 1869. A group of nuns visited his cabin in 1873 and found him '. . . resting

on a sofa which also answers for a bed, being constructed for the purpose – he showed all corners of his domicile – a table stand in the centre – glass case, barometers, thermometers, sword, gun, pistol – charts of all corners of the globe and articles too numerous to mention formed the furniture of the apartment which is called the "state room"'. He had an accident in Liverpool and was discharged from the ship in 1875, leaving Peter Robertson from Leith to take charge for the last few voyages as a steamship. Her first captain under sail alone, James Morris, left after a few weeks. His successor, Henry Stap, did not enjoy his time either, partly because the ship was unstable, and eventually he was obliged to abandon her in the Falkland Islands.

Officers

On the first voyage of 1845, Captain Hosken had six officers or mates to assist him, which probably allowed two on duty for each of three watches. By the time of the first Australian voyage in 1852, the ship usually had four mates – probably relieving the first mate from watchkeeping duties and having the other three stand one each. In July 1870 the fourth mate 'carelessly left his post and was down in the intermediate [deck] for not less than ¾ of an hour. The Captain happened to come on deck during his absence and of course missed him.'

Rosamond D'Ouseley describes the officers in 1869:

Mr Lawrenson, tall red whiskered and affectionately called by the sailors 'Ginger and Carrots' and by a friend of mine . . . the walking 'Firebrand'. 2nd Mr. Christie, good-looking and gentlemanly. 3rd Mr. Congdon, rough and black whiskers and beard being jolly too and the sailor's favourite, 4th Mr. Dusanteys, a very agreeable young person, extremely diminutive and pronounces his R's W's.

Lawrenson had spent 15 months in the Arctic during one of the many expeditions to search for the explorer John Franklin.

Felix Grenyer did not get on with the first

RIGHT Some of the officers of the ship, with Captain Mathews in the centre of the front row.
Standing, left to right:
Edward State, Cadet; Mr Cleave, 3rd Officer; Archibald Alexander, Sergeant; George Christian, Cadet; F.P. Smith, patentee of the screw propeller; Robert C. Lambert, Cadet; John Anjer, Purser; George M. Miller, Cadet; Chas Peters, 4th Officer.
Seated:
W. Martin, 2nd Officer; B.R. Mathews, Commander; H.T. Cox, 1st Officer. *(ss GB Trust)*

mate in 1867–68. 'Nobody liked the Chief Officer, he was anything but nice. Even the passengers hated him. . . .' He appeared to have prejudice against the Shetlanders who had been recruited by Captain Gray, and often 'cuffed' them. According to Grenyer:

I was having a wash just after breakfast, when suddenly the Boatswain came in and piped, 'All hands to furl sail'. I finished as soon as I could then rushed on deck, not much behind the others, and the Mate was standing by the scuttle, and as soon as I was clear of the scuttle he gave me a cuff and a push. I don't remember what he said, but that cuff and push put me in a blaze of wrath. So I took hold of him by the shoulders and pushed him backwards over the pinrail by the starboard fore rigging and walked aft as though nothing had happened.

Violence against the crew was not unusual on other voyages. 'The Capt. is very strict of course. His underlings take this cue accordingly. So far any remissness skulking on part of the crew & the fist or foot seems to be the usual application. . . .'

The captain had other ways of disciplining the crew, by stopping their allowance of drink, stopping their pay for all or part of the voyage and by locking them up. His right to flog men was doubtful by this time, but in 1853 a court martial was held on a sailor for 'insulting a lady and kicking one of the officers'. He was sentenced to 12 lashes and many of the gentlemen got up next morning to watch, but according to Anne Henning, '. . . I am glad to say they were disappointed for after he had been tied up to the mast and sufficiently frightened, he was let off at the intercession of Captain Snell. Of course this was all arranged with Captain Matthews beforehand and he was only brought up to save appearances with the sailors.'

The captain or officer of the watch did his duty within full view of crew and passengers. Under sail he would normally stand near the stern where he had a view of the sails and was close to the steering wheel and compass. There was nothing to stop him being interrupted by the first-class passengers promenading in the

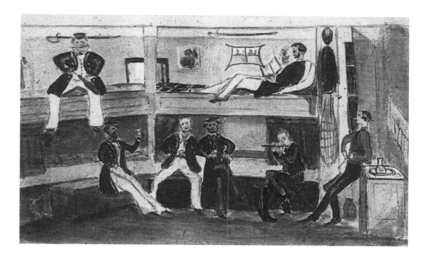

ABOVE 'Officers at Rest' by Moriss. *(ss GB Trust)*

area. Under steam he went up to the bridge where everyone could see him, and hear his orders shouted to the helmsman. Any indecision or panic would be apparent to the whole ship. In 1870, one passenger took it upon himself to mimic the officers.

It appears to me that he has a monomania for being Captain of the Ship. He will walk for hours directly underneath the Captain's

ABOVE View from the bridge. *(Jonathan Falconer)*

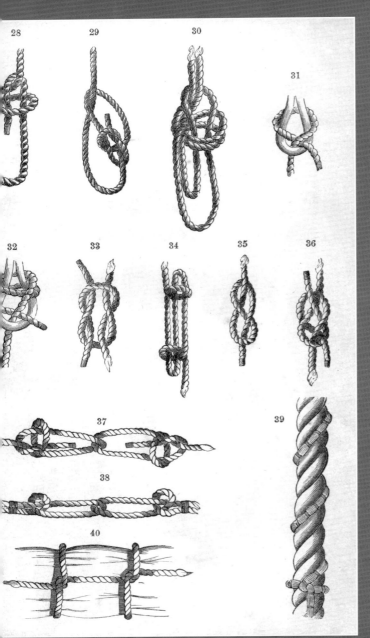

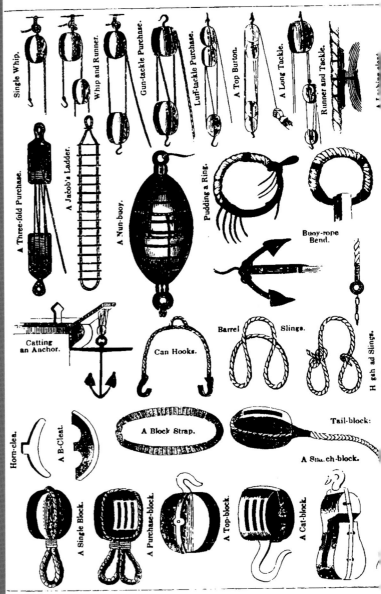

Cadets learn the ropes

Gibbs, Bright & Co. took on cadets to train as officers. According to instructions in 1854, 'Gentlemen by birth, our Cadets must be treated as such, and do not expect in the ordinary course to be called upon for menial service; for which purpose you are provided with common apprentices, or ordinary seamen.'

They were 'not to associate with the men before the mast' and were to be carefully trained, for the company was 'pledged to make them efficient seamen, and that they should be instructed in all parts of seamanship, such as reefing, furling, steering, use of lead-mark and deep sea soundings, sending up masts, yards, booms, stripping rigging, making and taking in sail, lowering boats, constructing rafts, stowing cargoes, baggage, stores &c.'

Their moral character required 'the greatest care', which was 'a duty always difficult, but peculiarly so in a passenger ship'.

Bridge mimicking their every possible way sometimes taking observations with one of these small eighteenpenny telescopes then waving his hand to an imaginary man at the wheel as though giving some very particular signs to that worthy for the necessary safety of the ship.

When one of John D. Parr's friends played a practical joke by mimicking the loud voice of the first mate, 'The Ship was in a roar of laughter for some time after this burst of wit.' Officers and midshipmen had uniforms, but when they took on the airs of naval officers they were not successful, '. . . all the officers and midshipmen attended, in their uniforms with swords on, which report says they did not very well know what to do with, but got them constantly hitched between their legs and among the benches . . .'.

Crew

The seamen of the British merchant marine saw many changes during the 40 years in which the *Great Britain* was in service. In 1843 the press gang was theoretically still empowered to force seamen into the Navy, though it had not been used since the last war in 1815. The Royal Navy and the merchant seamen were not different animals in 1843, but in the next decade the Navy began to train boys, sign them on for long service and issue them with uniforms, so by the 1880s the separation of the two types was almost complete.

To bridge the gap the government set up the Royal Naval Reserve for merchant seamen to train with the Navy, and a large proportion of the *Great Britain*'s men did so – Captain Gray was commissioned in it in 1862, and his relatively humble rank of lieutenant was often mentioned in advertisements. In the past British ships had been compelled to take on British crews under the Navigation Acts, but these were repealed by 1848. Legislators turned increasingly to safety and working conditions, notably in the famous act of 1876 which established the Plimsoll line.

The seaman had a distinctive character. In 1846, according to the *Nautical Magazine*, 'the generosity of the true seaman knows no bounds. . . . If he lacks a little polish, his

ingenuousness compensates. . . . His candour and sincerity are so universally acknowledged, as scarcely to need a passing remark. . . . His friendship is exemplified in the fraternal regard he entertains for his *shipmate*, and more especially his *messmate*.'

The seaman could be distinguished by his physique as well as his clothes. 'The remarkable stout thighs of seamen must strike the most careless observer. This arises from the constant practice of exercising the muscles in the action of going aloft.' Sailors were used to looking after their clothes, and according to Elizabeth Joseph, 'they wash quite as handy as any woman. They have pumps to get water – two men pumping while the others are washing.'

There was no strict segregation between crew and passengers and the young Thomas Park recorded: 'I was not many days on board ship before I was first favourite with the sailors, and was often to be found in the forecastle with them.' One of them made him a model ship, which he lost by trying to tow it astern.

RIGHT Merchant seaman's discharge papers for John Helm, who crewed on the *Great Britain* from 15 December 1864 to 22 May 1865. *(ss GB Trust)*

Reporting for duty

The crew was allocated duties at the start of the voyage. According to Dana, 'In dividing into watches, the master usually allows the officers to choose the men, one by one, alternately. . . . The men are divided as equally as possible, with reference to their qualities as able seamen, ordinary seamen, or boys. . . .'

A watch was a group of men who were on duty at a particular moment. It was also the period of four hours during which each watch was on duty, except for the dog watches, two hours each between four and eight, which gave variety to the daily routine.

Each seaman could expect to spend half his time on watch during the voyage, plus any time in which 'all hands' were called to carry out heavy work or in an emergency.

Maintenance, domestic and administrative staff worked during the day unless needed – they were known as 'idlers'. The seaman lived well forward in the ship, hence the expression 'before the mast' to describe the ordinary sailor.

BELOW The crew's day as divided into watches and rest periods. *(Author)*

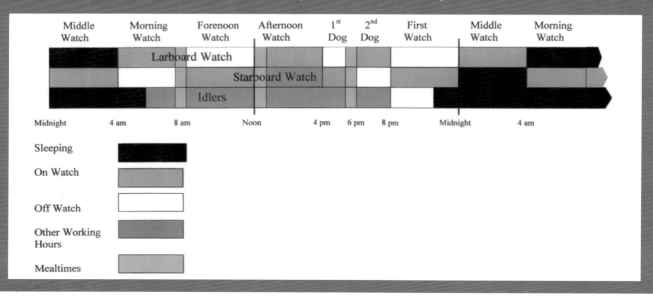

The best known member of the crew was Joseph Rogers, who had gained fame during the loss of the *Royal Charter* – another iron Gibbs, Bright ship built by William Patterson – in 1859. According to W.D. Waters, he was '. . . the very man that swam from the wreck at the risk of his own life with a Rope round his Wrist to the shore and was the cause of saveing many lives that would otherwise have gone to a Watery grave with the rest. He is a short dark set man very black hair and beard but the very type of an English sailor' – though he was actually Maltese.

For the first voyage in 1845 there were 4 helmsmen, 14 able and 6 ordinary seamen, 5 apprentices and a boy for duty on deck and in the rigging – a small number as the ship was expected to spend most of her time under steam, and the rigging was designed to be handled by as few as possible. When Felix Grenyer served as a seaman in 1867–68, there were 36 able and 12 ordinary seamen, stationed in pairs on each watch on the forecastle, fore rigging, fore yard, fore top, main rigging, main yard, main top, mizzen rigging and mizzen top – though they were not expected to confine their work to any particular area.

Competition was encouraged between masts. 'It was all hard work for us, but made light of with fun and excitement to try to finish our mast first. We had two Boatswain's Mates, besides the Boatswain, and they had to keep a sharp lookout to pipe "belay" in time, or the halliards might get parted.'

The male steerage passengers were

expected to help with unskilled work. John D. Parr reported, 'I have been going in for some nautical exercise pulling ropes running round the capstan + etc. I dare say in many instances more to the Sailors annoyance than otherwise.' But Felix Grenyer appreciated them. 'They were a great help when setting sail, they would tail on to the topsail halliards and with a walking chanty which all joined in with a will. . . .'

Anchors

As a steamship the *Great Britain* had less need for anchors in bad weather, as she could use her engines to prevent her from being driven on to a lee shore by the wind. When in port she mostly rode at anchor, partly because there were few docks large enough to hold her, and partly because it was far cheaper. In 1862, according to Mr Dearlove, the ship had

ABOVE A Trotman anchor in the *Great Britain* dry dock.
(Jonathan Falconer)

ABOVE Details of the Trotman anchor.
(Author)
1 *Arm lying flat*
2 *Lifting ring*
3 *Fluke shank*
4 *Stock*
5 *Ring for chain*
6 *Bolt*
7 *Arm and fluke buried in the ground*

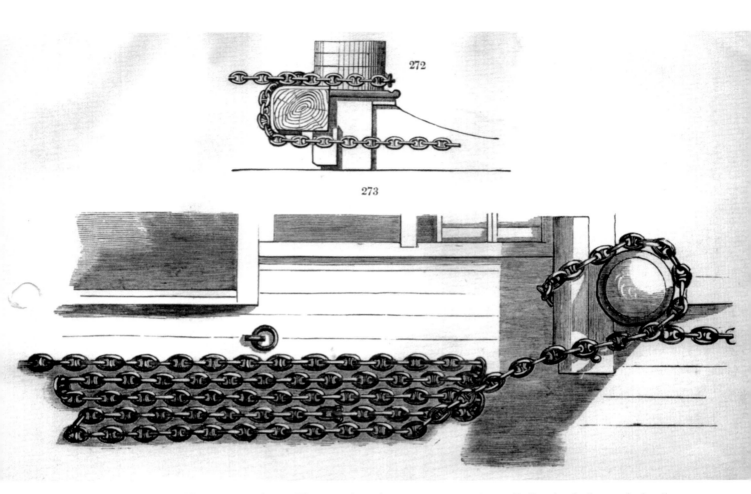

272

273

a 53cwt port anchor, a 50cwt starboard one
and a 47cwt spare, as well as an 11cwt stream
anchor which could be laid out by a boat.
The chain cable was 150 fathoms (900ft) long
and the links were 2¼in thick. The cable for
the stream anchor was the conventional 120
fathoms long and 1½in thick.

The *Great Britain* used Trotman anchors,
each with a bolt at the end of the shank.

*The peculiarity of the anchor is that the
arms pivot about this bolt, so that when
it takes hold the upper fluke is brought in*
*contact with the shank, thus reducing the
height above ground, and rendering it almost
impossible for the cable to get entangled
round it, or, in other words, for the anchor to
become foul.*

It did well in tests conducted by a
government committee in 1852, with the best
holding-power of six types tried. Chain cable
was used on all modern ships. It was less likely
to rot, it was stronger, and it lay flat on the sea
bottom and therefore it pulled on the anchor at
a better angle than rope.

Approaching port, the anchor was made
ready to drop and a suitable length of chain was
laid out on the deck. The ship headed slowly
into the tide or current, then stopped. She
should then release the anchor, paying the cable
out slowly as the ship was driven backwards. As
Edward Towle wrote in 1852, 'the noise of the
enormous chain cable to which [the anchor] was
fastened rushing through the hawse hole was
without exaggeration like thunder'.

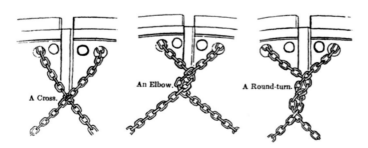

A Cross.   An Elbow.   A Round-turn.

LEFT The cathead, which acted like a crane to raise the anchor. (*Jonathan Falconer*)

BELOW LEFT *Great Britain*'s capstan. (*Jonathan Falconer*)

BELOW The windlass. (*Jonathan Falconer*)

TOP The wheels of the windlass had grooves designed to take the links of the anchor chains.

ABOVE RIGHT The ends of the anchor chains are fitted to ropes when not in use.

ABOVE Stoppers were screwed down to fix the anchor chain when it was in use.

RIGHT A catted anchor on a warship of the period, of the conventional pattern, not the Trotman.

A ship normally let out a length of cable equal to three times the depth of water – 120ft of cable if the water was 40ft deep. This meant that she would swing with the tide in a circle with that radius. Other ships in the anchorage would probably swing the same way, but the officers had to make sure that she would not hit any fixed obstruction while turning. Anchoring bow and stern was highly dangerous as the ship would not necessarily point into the wind in a gale, but the circle could be minimised by using two anchors from the bow – though there was the danger that they would become tangled with each turn of the tide, creating a 'foul hawse' which was very difficult to clear. The *Great Britain* normally used a single anchor in the Mersey and at Melbourne.

On sailing from the Mersey in October 1861, according to the ship's newspaper, the sails were let loose ready to be set when required, then 'the sailors sang merrily as they danced round the capstan, weighing the ponderous anchor which lay deeply embedded in the Mersey. The tide at this time was nearly full, and the captain was anxious to get away while there was sufficiency of water on the bar. . . .' But there was a delay, and the ship had to anchor again until the next morning.

'The first sound that greeted our ears was the merry and lively chant of the sailors heaving at the capstan, to weigh anchor; and at 8 a.m. the ponderous cable was all aboard, the signal guns were fired, and the "Great Britain" started her voyage. . . .' The ship would certainly have steam up before raising the anchor, and would usually leave port under power; though in March 1870 the wind was suitable to come out under sail.

The capstan could be dangerous. Levers, known as pawls, were supposed to control the motion, but they did not always work, as Grenyer reported:

. . . a sudden heavy gust of wind came from ahead and gave us stern way. All the hands were at the capstan and we could not hold her, and before anyone could shove the pawls in, the capstan took charge, most of us dropped down clear of the bars, as they were spinning round at a tremendous rate, but 3 men got struck by them and hurt badly. . . .

Once out of the water, the anchor was hauled up on the cathead, a wooden crane projecting from the bow, to be 'catted' or hung vertically. Then it had to be 'fished' by raising its crown and levelling the shank so that it would not move about too much in bad weather. This could be hard work with the Trotman's, for 'With the common anchor they could catch either one side or the other; but with the Trotman's the hook must go into a large ring bolt close to the shank; and it would be a very difficult operation with a heavy sea to effect that manoeuvre.'

The officer of the watch under sail

Captain Gray preferred sail to steam and in 1862 he remarked that fine weather 'may be very comfortable to passengers, but not to me. As a sailor I delight in "the wet sheet and flowing main".'

The officer of the watch had to decide the amount of sail to be set, calling the captain if a major change had to be made or danger threatened. He needed a good eye for the weather. In calm or light winds he looked out for increases in strength which might help him. He would spot any oncoming storm by clouds or disturbance on the sea, but not always successfully, according to Rosamond D'Ouseley:

Friday – beautiful morning before & after breakfast but gradually in the West dark clouds coming up (the wind having changed) which the Captain & many of the passengers knew to be a squall & immediately we heard orders to haul in the sails, but before they were all in – one or two were torn to ribbons & flapping against the sides of the ship made a noise like thunder, while the sea began to rise higher & higher until some of the waves came on deck.

As the wind got up he noticed any increase in the heel of the ship, which would reduce the effect of the sails, so it was time to take in some sail. He also had to look out for the balance of the rig. Too much sail set forward and the ship would be pushed off the wind, or have lee helm; too much aft and it would have weather helm. He also had to make sure that the sails were set at the right angle to the wind for the greatest effect, seeing that the wind was passing smoothly over both sides. A fore-and-aft sail would tend to rustle near its forward edge, where the airflow was interrupted by the mast. When this became excessive, it was time to make adjustments.

The theory of sailing

Like nearly all large ships of the time, the *Great Britain* carried both fore-and-aft and square sails, and the proportions of the two varied considerably over 40 years of service. The square sail is not literally square, but its top and bottom edges – the head and foot – are parallel. It is hung from a yard in front of the mast and in its neutral position it is at right angles to the centre line of the ship. The fore-and-aft sail, as its name implies, is in line with the centre line in its neutral position (when it is not likely to be in use). On a modern yacht it is triangular in shape, but in the 19th century it was more likely to be an irregular quadrilateral.

Though the theory was not known at the time, every sail operated like an aeroplane wing,

set in the vertical rather than horizontal plane. It assumed a curved section and wind moved faster on the outermost side and created less dense air there, which tended to pull it forward. A component of this force (known as leeway) was pushing the ship sideways while a stronger component was pushing the ship forwards. The art of the ship designer was to create a vessel in which leeway was minimised, while resistance to the forward motion was the least possible. At the same time the ship had to be wide and stable enough to resist the heeling created by having the wind on one side. The sail operated differently if the wind was directly behind it, like a stalled aerofoil or a parachute, which made it less efficient.

The square sail was at its best when the wind was coming over the side of the ship, or astern of that. Because of its position forward of the mast, it could only be swung round to a limited extent before it met the shrouds, so it was only expected to sail within six points, or 67½ degrees, of the wind; a fore-and-aft rigged ship could sail within four points or 45 degrees. The square sail needed a great deal of work, but it was better for large ships, for it could be divided horizontally into separate sails, each handled individually. Each yard could be set at a slightly different angle, which was important as the wind was stronger and coming from a slightly different direction higher up.

Fore-and-aft sails

The fore-and-aft sail was much easier to handle. Its head was attached to a gaff at an angle to the mast, which was held in position by ropes known as halyards. In its first incarnation, the *Great Britain* used the standing gaff, which was kept aloft all the time, and the sail was furled by hauling it towards the mast by ropes known as brails.

When setting a sail, it was necessary to make sure the peak of the gaff was on the right side of the topmast stay so the peak halyard might have to be loosened to pass the gaff under the stay. The brails were loosened and the vangs which controlled the top end of the gaff were adjusted so that it was pointing as close into the wind as possible. The head of the sail was attached to the gaff by hoops and it was pulled out along it by the outhauler. The rope known as the sheet was used to control the bottom corner of the sail, allowing it to flap until it was fully set. The sail was loose-footed, in that there was no boom to stretch its foot. The sheet was hauled in until the wind was passing evenly over the sail and the vang was adjusted to match, so that the sail had begun its work.

When furling, the process was reversed. The sheet was eased out and the sail was pointed into the wind. A rope known as the inhauler was used to pull the head of the sail along the gaff towards the mast, while brails attached to the after edge of the sail, the leech, helped to furl the rest of it.

In later versions of the rig, the feet of the gaff sails were spread by a boom roughly parallel to the deck. When furling or reefing the gaff was simply let down and gravity did most of the hard work, while the furling was done at deck level and there was rarely any need for men to go aloft. Of course, it took some effort to hoist the gaff in the first place, but that could be helped by winches. It was rarely necessary to raise it in an emergency, so a fore-and-aft rigged ship did not need a large number of men standing by in case of a sudden storm.

The main disadvantage of the gaff sail was that it could not be divided, so there was relatively little control over its shape. When filled with wind it would tend to curve in the vertical as well as the horizontal plane, so most of it was not at an ideal angle to the wind.

The vang controlled the outer end of the gaff, but it ended at the deck so it was at an awkward angle and had little purchase. Because of this lack of control it was dangerous to make the sails too tall, as they might catch the strong upper winds at the wrong angle and tend to capsize the ship – this is why the original masts of the *Great Britain* seemed unduly short, and disappointed observers in New York.

The *Great Britain* also used gaff topsails which were set in the angle between the gaff and the head of the mast. Each one seems to have been spread by a small gaff, and it was loose-footed, so that the forward lower corner, the tack, was pulled slightly forward of the mast by a rope leading to the deck, to extend the

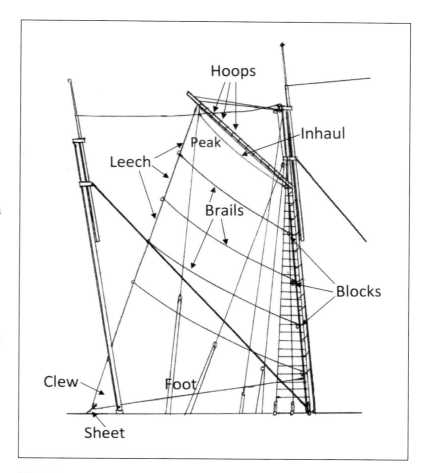

ABOVE **The same mast with the sail set.** *(Author)*

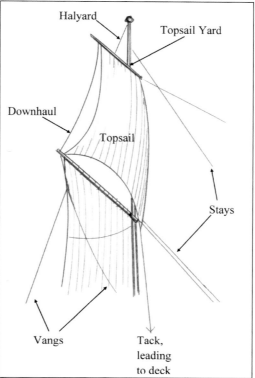

LEFT **A gaff topsail and the ropes attached to it.** *(Author)*

RIGHT The pin rails at the foot of one of the masts, used for securing ropes. *(Jonathan Falconer)*

BELOW The belaying points for the rigging of the fore mast. *(Author)*

BELOW RIGHT Belaying points for a typical after mast. *(Author)*

area and improve the airflow. They too could be set and lowered from the deck, and they followed the movements of the gaff sail. Since the topmasts were not well supported, these sails were probably only used in very light winds.

Like nearly all ships of the time, the *Great Britain* also carried staysails, which were mostly triangular with the longest edge set on the stays – the rigging lines leading diagonally forward from the masts. The most prominent were the jib sails, set between the bowsprit and the foremast.

Square sail

A square sail needed many more ropes than a fore-and-aft sail. If its angle to the wind was to be altered by means of sheets and braces, it needed men letting off on one

side and hauling on the other. If five sails were set on a mast, as with the *Great Britain* in her Australia days, the sailors would have to take in hand the braces which controlled both ends of the royals, the topgallant yards, the upper and lower topsail yards and the course, as well as the sheet which controlled the lower corners of the course. The ropes were manipulated until the officer called out 'well', then hauled taut and the ends were belayed or made fast to pins along the sides.

A square sail was set by sending men aloft to loose the ropes known as gaskets which held it to the yard. Men on deck hauled on the sheets to bring the corners of the sail close to the yard below. The upper yards then had to be raised to expose the maximum area of sail to the wind. If the wind got up suddenly it was possible to take some of the wind out of the sail by lowering the yard, then hauling the sail up to the yard; but men still had to be sent aloft to furl it with gaskets, or it would flap about uncontrollably. It was also possible to reef some of the sails, to take in part of them by tying reef points along the sail together across the yard. This could be very labour intensive, especially in the rig of 1852 and before the topsails were divided in two in 1857 – it was claimed that up to 50 men were needed on a yard for reefing.

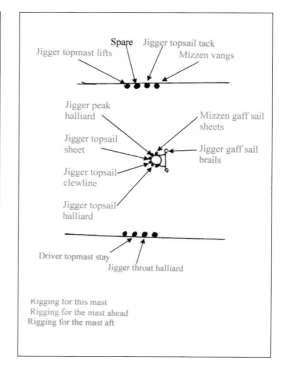

Fore topmast lifts
Main tack
Spare
Main lazy tack
Fore topsail tack
Fore staysails sheet
Jib sheet
Flying jib sheet
Fore peak halliard
Downhauls for headsails
Flying and fore jib halliards
Fore topsail sheet
Fore gaff sail brails
Fore topsail clewline
Fore staysail halliards
Bowsprit
Fore topsail halliard
Main topmast stay
Fore throat halliard

Ropes are only show on the starboard side if they differ from those on the port side

Headsail ropes
Foremast sails ropes
Mainmast sails

Spare
Jigger topmast lifts
Jigger topsail tack
Mizzen vangs
Jigger peak halliard
Mizzen gaff sail sheets
Jigger topsail sheet
Jigger gaff sail brails
Jigger topsail clewline
Jigger topsail halliard
Driver topmast stay
Jigger throat halliard

Rigging for this mast
Rigging for the mast ahead
Rigging for the mast aft

There was always work for the sailor under sail, as described by Rosamond D'Ouseley in 1869. 'I am sitting at the side of the ship from the deck leaning against a large coil of rope, sailors are continually changing sails, hauling ropes & they all sing a stave, then pull together; the most popular one is "Away, haul away; Joe" (pull).'

The points of sailing

If the wind was less than 30 degrees off the bow, the ship was head to wind and there was no point in setting sail at all – indeed square sails would back and retard the motion. Between 30 and 45 degrees, a fore-and-aft sail might steady the ship to prevent rolling, but it did nothing to increase the power. After 45 degrees (67 degrees for square sails) the sails would begin to contribute to the ship's progress and she was said to be close-hauled. If the helmsman came a point or two further from the wind direction, or 'bore away', then the ship was 'going free' and the sails were approaching their maximum efficiency. A little further and the wind was at right angles to the ship and 'on the beam'. After that, she was running before the wind which was behind and the sails tended to mask one another, as one passenger recorded in 1863, 'Wind gets round to the stern again which retards our speed as all sails cannot get filled, one catching it from another.'

Following the ancient practice with the tiller, it was normal for a ship to steer with the top edge of the wheel pointing away from the direction to be moved – the opposite of modern practice. The helmsman could expect a gap of several seconds before the bows started to move round, though that was small in the *Great Britain*. As he approached the course he wanted, he might begin to take turns off the wheel and perhaps put it slightly in the opposite direction, as the ship would continue to turn unless compensated. In a tight corner he would obey the officer's orders such as 'hard a port' and 'steady as she goes'.

Apart from that there were three ways of steering a sailing ship. In harbour or close to land, the helmsman might point at a fixed object on shore, such as a lighthouse or church steeple. This did not allow for the ship being swept sideways by the tide, so it was better to keep two objects in line, like the marks at Bootle in the entrance to the Mersey. At sea, a ship was usually steered by a compass course as given by the navigating officer. And if the wind was unfavourable, the helmsman was

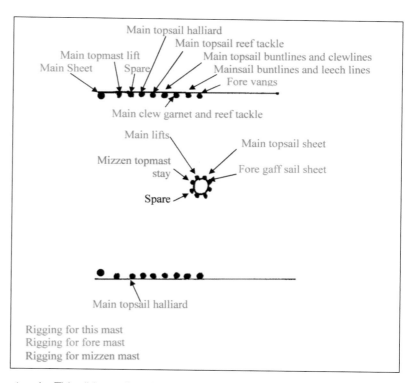

ABOVE **Mainmast belaying points.** *(Author)*

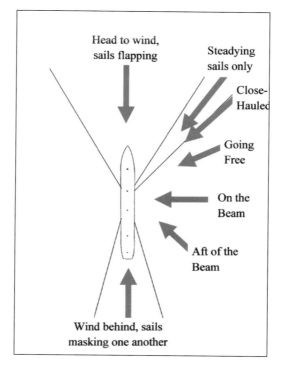

LEFT **The points of sailing on a fore-and-aft rigged ship.** *(Author)*

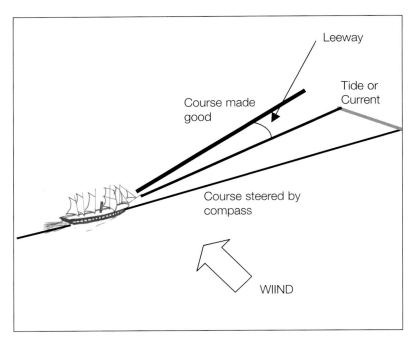

ordered to steer 'by the wind', getting the ship as close as possible to the direction of the wind while keeping the sails filled – 'full and by'. That was less likely in the *Great Britain* as she would probably use her engines when the wind was contrary. The other two, by objects on land and by the compass, were common under steam and under sail.

Steaming

On the face of it, life was far easier for the officer of the watch and the seamen when the ship was steaming. After furling the sails the seamen might be employed in cleaning or light maintenance duties, as Elizabeth Joseph noted in 1870: 'The sailors are making ropes. They are not idle a moment.' In good visibility the officer of the watch only had to keep the ship on course and make sure that the lookouts were alert. But in case of fog and the presence of danger it became much more difficult. Well out in the North Atlantic in 1862, according to Mr Dearlove,

. . . one of the passengers saw a vessel right ahead of us just in our track and going the same way. He saw it before our look out the night was hazy they appear to have seen us a minute or two before we saw them almost immediately set a Tar Barrel or some Spirits of Turpentine in a Blaze – 5 or 6 men rushed to our wheel and put it hard over and we went off nearly at a right angle and passed about 100 yards on her Starboard side. The voices from the Brig could be distinctly heard aboard us another few minutes and we should have run her down and most

ABOVE The ship among icebergs near Cape Horn. *(Illustrated London News)*

likely every soul on board of her would have perished as being brig a small brig the "Great Britain" at the speed she was going I think she would scarcely felt the collision especially as she would have run into the Brig's stern.

Collisions were far more likely in the confined waters of the Irish Sea and the approaches to Liverpool. In April 1862 the captain asked some of the passengers to cease their late-night revelries because the noise made it difficult to hear other vessels in the fog in which it was 'almost impossible to see a yard ahead of the vessel'. The engines had to be stopped six times that night to prevent collisions.

Captain Hosken largely ignored the danger of ice off Newfoundland in 1845, perhaps because Claxton had claimed that among ice floes, 'Iron sailing vessels, and steamers fitted with screws, may fearlessly keep up their speed and continue on their course.' Full-size icebergs were also encountered. In 1870 Captain Gray posted three mates and four seamen as lookouts when south of the Cape; 83 were spotted in one day.

The officer of the watch also had to be ready for emergency, for example in 1864, 'Aroused . . . with that awful and indescribable cry, "A man overboard!" All was instantaneous confusion. I ran aft and came in contact with Mr. Angus, the fourth officer, rushing towards the engine telegraph to stop her.'

Perhaps the most important duty of the officer of the watch under steam was to look out for changing conditions, and to set some sail, or perhaps raise the screw. Passenger John Gurner believed Captain Gray knew 'when to raise steam and when to rely upon sail power'. But he did not always get it right according to the impatient Mr Dearlove:

At noon we had no wind and the Engine was stopped and we was not going 2 miles an hour. Wind during afternoon did freshen a little moving ship at 3 or 4 knots an hour. I am surprised the steam is not kept up, perhaps because they are frightened of getting short of coal. What folly not to use them when we want them and do without if it is a requisite nearer home perhaps keep coal which we may never want.

Probably no ships used combined sail and steam more than the *Great Britain* in her heyday. On her voyage to Australia in 1860 she was under sail alone for 52.8% of the time, steam alone for 15.5%, and combined sail and steam for 31.7%.

The seaman using any kind of sail always has to be aware of the difference between true and apparent wind. If the ship was sailing at 10 knots, a true wind at 45 degrees over the bow would seem like 34 degrees in a 10-knot wind. This was more important when using steam assisted by sail, as the motion created by the engines would create its own wind. The apparent wind would draw further ahead and would be less use; or if it was behind the ship, it would have to be considerably stronger than the ship's own speed to be of any use. There was only a sector of about 45 degrees on each side of the ship where the wind could be useful. In many cases it must have seemed far easier to keep the sails furled and carry on, but the fact that the *Great Britain* used her sails so much is an indication of how the officers felt it necessary to get the best out of them, and to save coal.

Chapter Four

Navigating the *Great Britain*

Navigation is a vital skill for the officers of any ocean-going vessel and is used to determine a ship's position and course when at sea. Out of sight of land for days and weeks, the SS *Great Britain*'s navigators would have used a combination of dead reckoning and celestial navigation, aided by a sextant, nautical charts and tide tables to assist the ship's passage across the ocean.

OPPOSITE The tools of the trade: parallel rules, a telescope and a quadrant on top of a log book and a chart. *(Jonathan Falconer)*

81

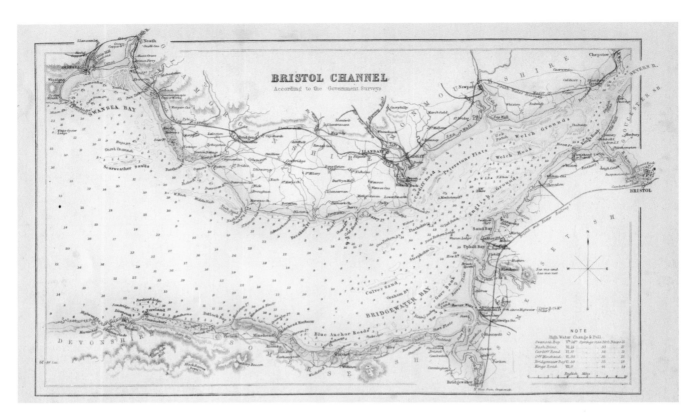

BRISTOL CHANNEL
According to the Government Surveys

ABOVE Navigational chart of the Bristol Channel from 1859, showing depth soundings and a variety of different navigational features.

Navigation and pilotage

Navigation comprised three main classes: pilotage, coastal and ocean navigation. Pilotage was used close to a port, or within its confined waters, and was usually done by a qualified local pilot. Coastal navigation was needed at the beginning and end of a voyage. It was used for a passage along a coast, or crossing a relatively narrow sea such as the English Channel or Irish Sea. The voyage would be mostly in sight of land, other than perhaps for a few hours or in fog or poor visibility. Ocean navigation involved days and weeks out of sight of land.

Tides are especially important in the waters round the British Isles where they are very strong. A tide has two effects. It is a large movement of water, and the navigator had to know its strength and whether it was with or against him, or if it was a cross tide which would sweep him well off his course unless he adjusted for it. He used tide tables to find out the speed and direction at that moment, and drew a triangle in which one side was the distance and course steered, another was the strength and direction of the tide and the third was actual distance and direction travelled.

Tides could also cause a problem in harbour. After the first arrival in New York, the *Great Britain* tried for an hour to get alongside Clinton Street Pier, 'but in vain, in consequence of the strong ebb tide, and the effort was given up until slack water'. The other effect of the tide was to raise and lower the level of the water, which made it essential to choose carefully the time of entry to a port such as Liverpool or Bristol.

The *Great Britain* was active during a golden age of navigational science. Scientific naval officers such as Matthew Maury and Admiral Robert Fitzroy were producing studies of tides, currents and climate. Many new lighthouses had been erected in the last few decades, and some were fitted with the fresnel lens which projected a light for up to 20 miles.

By the 1840s much of the world was covered by accurate and often beautiful nautical charts, mainly produced by the Hydrographic Department of the Admiralty under Francis Beaufort. To represent a sphere on a flat surface, charts were usually done on Mercator's projection, in which the world was 'opened out' towards the poles. A minute of latitude was equal to one nautical mile and the navigator could use his dividers to calculate the distances,

but on a small-scale chart he had to be careful to take the measurements from the latitude he was in, as it varied from north to south.

A chart included a compass rose with directions marked on it, and when plotting a course the navigator used his parallel rules to line it up. The chart had contour lines with the rough outline of the depth of water in various places, and for more detail it had depths marked in fathoms and feet in spots which the survey ship had plotted. In areas that were dry at low tide, the numbers were underlined and represented the height above chart datum. That was set at the level of 'low water ordinary spring tides', which was almost the lowest the navigator would expect to encounter.

The chart often included coastal views to identify features on land, but it was sparing with the land detail, while being meticulous with the positions of coastlines, rocks, shoals and other sea features. It recorded the nature of the bottom, whether sand, gravel, rock, mud and so on. This could be useful when anchoring, and a sample could give some indication of position.

Other essential publications included tide tables for the ports to be visited, pilot books which gave verbal descriptions to supplement charts, and the monthly *Notices to Mariners* which were published by authority of the Admiralty, in the *Nautical Magazine.* There were lists of lighthouses, and almanacs which gave the positions of sun and stars at any moment during the year.

In shallow water, the most important instrument was the lead line, which was used to find the depth of water under the ship. The hand lead was used in rivers and harbours, the larger deep-sea or 'dipsey' lead in depths of up to 110 fathoms in the open sea. The lead had to be cast forward of the ship to allow

ABOVE New York's East River in 1837, as seen from Brooklyn. *(TopFoto)*

BELOW Dividers for measuring distance on a chart. The curves in the upper part made it easier to open and close them single-handed. *(Author)*

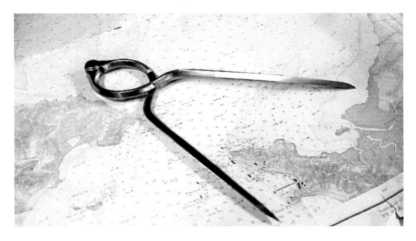

time for it to sink as the ship moved forward. It was skilled work and one of the benchmarks of an able seaman. The lead had a hollow in the base filled with tallow to pick up a sample of the sea bottom. The navigator could use it to take 'soundings' of the shallower area which indicated whether he was approaching the continental shelf round the British Isles. He could also use it to decide where to anchor.

Liverpool

The *Great Britain* did not return to Bristol until 1970, and her home port for the rest of her career was Liverpool. Apart from the problem with the size of the lock, the directors of the GWSS were already having difficulty with the attitude of the Bristol Docks authority and had had several differences about payment.

RIGHT A view of Liverpool and the river Mersey, 1886. *(TopFoto)*

After April 1843 the *Great Western* operated from Liverpool and only went to Bristol for annual maintenance.

Although Liverpool was further from London, it was linked by rail from 1838 and the journey took about six hours instead of three to Bristol, which was insignificant in the context of an ocean voyage. Liverpool had immediate access to the industrial heartlands of Lancashire, and had excellent communications with Ireland. Ships operating from Liverpool also had the choice of passing round the north or south of Ireland as circumstances demanded.

Over the next 40 years, Liverpool grew from a population of 286,000 in 1841 to 553,000 in 1881. It was handling nearly 8 million tons of shipping per year by the 1880s, had built 22 more docks (not including those across the river at Birkenhead) and had become the greatest port in the world. The port would see off more than 4 million emigrants from 1853 to 1882, though less than 4% of these were for Australia and New Zealand. Liverpool's merchants and ship-owners tried hard to raise the cultural standards of the city by founding art galleries and a university, but there was a constantly shifting population of seamen, emigrants and those who tried to exploit them, which made it a lively place, to say the least.

Liverpool was not as fiercely tidal as Bristol, having a maximum range of 31½ft compared with more than 40ft, but as a pilot book of 1839 put it, '. . .its entrances are extremely difficult and intricate, being obstructed by numerous and extensive sand-banks, which stretch to the northward and eastward of the Rivers Dee and Mersey. . .'. The sandbanks were constantly shifting and attempts to dredge them had failed, so pilots were compulsory for ships entering the port. The *Great Britain*'s favourite pilot was Henry Ennis, who had been born in 1827 and became fully qualified in 1849. He had already piloted the ship twice before he conducted her trials after the refit in 1852.

We think it due to Mr Henry Ennis pilot to express our satisfaction in the way in which he has always conducted himself while having the 'Great Britain' (SS) under his charge. . . . His carefulness and attention to the safety of the ship has been so marked that we think it right to bring the same to your notice.

ABOVE The *Great Britain* leaving Liverpool in 1853. Her sails are backing, perhaps to slow her down as the pilot is taken off. *(ss GB Trust)*

BELOW Tides are still strong in the Mersey, as shown in this photograph of the Mersey Ferry crossing the river. *(TopFoto)*

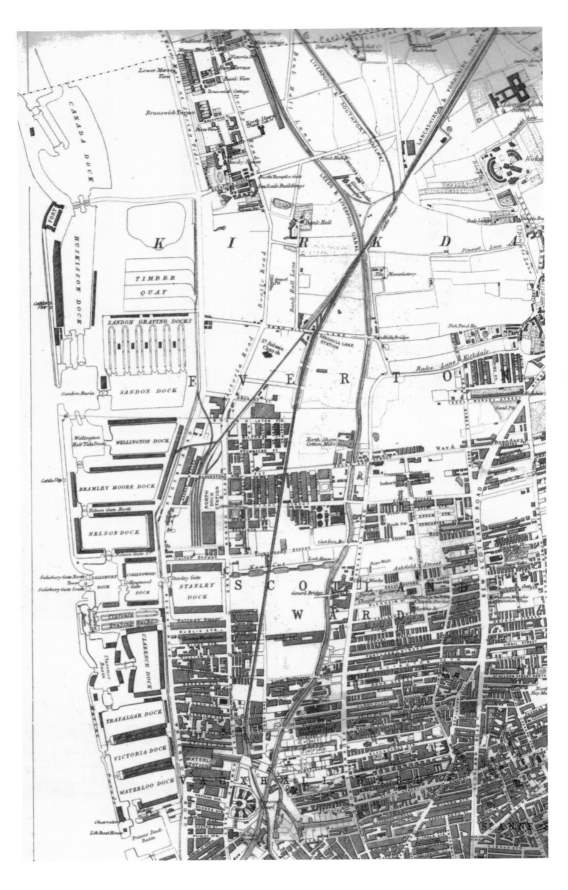

Inward bound, the pilot was usually picked up off Port Lynas on the north-east corner of Anglesea, which was always a welcome sight for the travellers. Victoria Channel was the best route in the 1840s, and it could be entered three hours earlier in the tidal range than any of the other routes.

The entrance was marked by a floating bell beacon. The navigator brought this into line with the Formby Light Vessel and the Crosby Lighthouse, which was 96ft above half-tide level. He should pass close to the beacon then steer South-East by East ¾ East for three-quarters of a mile, then between the red and black buoys marking the edges of the channel – though he was warned that 'Running for these channels in hard gales, the buoys are not much to be depended on, as they are then covered with spray and spoon-drift.' He would pass over the steep edge of a bar with only 11ft of water at low tide. After three-quarters of a mile he could pass either side of the Formby Light Vessel at the junction of the Formby and Crosby Channels.

He had to steer South-South-East ¾ East for 2¾ miles to the Crosby Light Vessel, leaving plenty of space between him and the North Burbo Bank, then set a course of South by East ¼ East, 'which will lead you fairly into the Mersey'.

On the Mersey

Inside the port the tides could be strong – up to 7 knots off the town. The *Great Britain* usually anchored in the only safe area: 'In Sloyne Roads and Bombro Pool, which comprehends the space from Tranmere Ferry to a little above Beddington Creek, is very good anchorage, with all winds, but the nearer the ferry the better, as here the tide does not set with that velocity as in the stream, and the ground is good.'

The river could be busy, as recorded by the American writer Nathaniel Hawthorne during a visit to the *Great Britain* in 1853:

There are a great many steamers, plying up and down the river, to various landings in the vicinity; and a good many steam tugs; also, many boats . . . also, here and there a yacht or pleasure boat; also, a few ships riding stately at their anchors, and probably on the point of sailing. . . . The small black steamers, whizzing industriously along, make up the chief life of the scene.

Rather than sealing off part of the river as at Bristol, the Liverpudlians dug out wet docks

Safety first on the Mersey

The *Great Britain* had to obey the rules of the road in a crowded waterway. Those for sailing ships were largely based on the direction of the wind and point of sailing. Steamships were expected to keep out of the way of sailing ships, though a great liner was not likely to manoeuvre sharply to avoid a tiny Mersey flat or a yacht.

When one ship met another nearly end on, both would put their helms to port. An overtaking vessel was responsible for keeping clear of the other; and when two steamships were crossing, 'the ship which has the other on her starboard shall keep out of the way of the other'.

In fact, the *Great Britain*'s only collisions at Liverpool were when she was at anchor in 1870 and a schooner got tangled up with her cable, and four years later when she collided with the *Mysore* in the Sloyne.

BELOW Landing stage on the Mersey. *(Author)*

which were for loading and unloading in a safe and tide-free environment – they had opened 18 by 1845. The *Great Britain* used the Coburg Dock of 1840 during her first visit in 1845, but mainly so that she could be opened to visitors.

The *Great Britain* did not need tugs to take her into harbour against contrary winds (except after her propeller had lost all its blades in 1845). Even with a hastily repaired propeller shaft off Melbourne in 1868, Captain Gray declined their use. But Liverpool tugs had many other functions. On her maiden voyage to New York, the tug *Dreadnought* took off some of the distinguished guests. During the first Australian voyage in 1852 a tug left the Seacombe Slip to bring the passengers on board.

The *Great Britain* could operate without wet docks, but dry docks were vital for repair. Her size was not so much of a problem at Liverpool. Her breadth of 50ft was no longer exceptional, and it was quite common to build dry docks to hold more than one ship, with one behind another. Queen's Graving Dock No. 2 was 70ft wide and 467ft long. The main problem was the entry through the Queen's Basin, for she had to turn her 322ft length through 90 degrees in a space that was only about 340ft wide. It seems likely that she was manoeuvred by means of ropes on the shore – all docks were provided with numerous bollards and capstans for attaching ropes. It was her draught, however, which caused the most concern. The *Liverpool Courier* reported:

There being a very light spring tide on Saturday, a very short draft of water was necessary in order to get the leviathan steamer safely into the graving dock. After every exertion, by trimming coals, cables, anchors, etc. at seven o'clock Saturday morning the ship was drawing 14ft 8ins aft and 12ft 6ins forward. Captain Hosken, seeing this, immediately ordered a bolt to be taken out of the bow and allowed the foremost compartment to fill with water. This in a short time brought the ship to 14ft aft and 13ft 1in forward, putting her in the trim required, and at the same time, affording satisfactory proof of the utility and efficiency of the water-tight bulkheads.

She often used the Sandon Docks after they were built in 1851. The basin was much wider than Queen's, and it gave access to six dry docks, all 565ft long and three of them at least 60ft wide.

New York

New York was a great natural harbour where the rise and fall of the tide was rarely more than 5ft. The area could be identified from many miles off by the New Jersey mountains known to seamen as 'Never-Sink' – though not in the 1852 voyage when the area was covered in fog.

The ship passed through the Gedney Channel to reach the north end of Sandy Hook, a 5-mile spit that marked the entrance to the harbour. She picked up a pilot from one of the highly competitive cutters, then went south of a light vessel which marked the edge of a series of sandbanks. She passed through the Narrows between Long Island and Staten Island, and after into Upper Bay, the main anchorage, which was 4 miles wide. Visitors often remarked on the beautiful countryside, though the famous New York skyline was a long way in the future.

The ship headed for the East River side of Manhattan. New York did not have enclosed docks like the British ports, but large numbers of quays (known as slips or wharves) at right angles to the shoreline – Clinton Street Pier in the East River had been extended to cater for her.

BELOW Liverpool's docks extended for nine miles along the banks of the Mersey. They are seen here in the 1890s. In the foreground is one of the great steel ocean windjammers which, like the *Great Britain*, carried heavy and non-perishable articles such as grain and wool. *(TopFoto)*

Melbourne

Melbourne was situated on another great natural harbour, Port Phillip, which had 800 square miles of sheltered water, three-quarters of it suitable for anchorage. The town had been founded in 1837, and had trebled to around 100,000 people during the gold rush.

On approaching from the west, the seamen would look out for the lighthouse on Cape Otway, a circular white tower 300ft above the water with a bright flash once a minute. They were now 60 miles from the entrance to the harbour and would keep 4 miles south of the headland, 'skirting a coast peculiar by its undulating character and its dense forest of trees almost destitute of foliage'. Soon they would sight Arthur's Seat, a mountain to the east of the entrance, and then Flinders Point, 'an isolated, saddle shaped scrubby hill about 150 feet high', which was 'an excellent mark for the port'.

Between 3 and 12 miles from the entrance they might find one of the pilot cutters, as the ship's newspaper reported in 1861, '. . . shortly after sundown took our pilot on board amid a round of cheers which told of the joy of all at the happy termination of our voyage'.

The tides could run at up to 7 knots across the entrance, and they were well advised to wait for a suitable moment to enter. They could find the state of the tide by watching flag signals from Point Lonsdale to the west of the entrance – this was better than consulting tables, for

the level of the water was much affected by the wind. Passing through the narrow channel between Point Lonsdale and Point Nepean, they would take care to avoid Corsair Rock, 11ft under the water level, and Point Lonsdale Reef, which ran out three cables or 600yd from the point. The ship stopped off at 'the pretty hamlet of Queenscliff' to take on the health officers and customs officials. Once in, they followed the West Channel for about 30 miles to anchor in Hobson's Bay, often among a great mass of shipping in the gold rush days. Passengers could be landed at one of the piers in the bay, or go up the Yarra River to the town of Melbourne.

ABOVE A bird's-eye view of the city of New York with Battery Park in the foreground and Brooklyn Heights at lower right, c.1856. It shows the numerous piers on both sides of Manhattan, with Clinton Street on the right. *(TopFoto)*

LEFT Queenscliff at the entrance to the natural harbour of Port Phillip, where Melbourne is situated. *(TopFoto)*

Dead reckoning

One of the navigator's problems was that the ship was rarely travelling in the same direction as her head was pointing. In all circumstances the basic navigational technique was dead reckoning, keeping track of the ship's position on a chart by the hour or by the day. The courses steered were logged (and these might vary considerably if it was necessary to catch a wind or avoid a hazard). Speed was measured in knots or nautical miles per hour, each equal to 1.15 land miles; it was found by trailing a log over the stern and measuring the number of knots in its rope as it passed through the navigator's fingers in a given time.

It was also possible to use a patent log, such as the one invented by Edward Massey around 1840. This had a propeller-like shape and was trailed over the stern to give a continuous reading, as well as a measure of the distance run. Guppy was sceptical about its value in 1845: 'When a vessel was going less than 5 knots, it did not indicate sufficient distance; between 5 knots and 6 knots, it was very correct; but above that speed, it again gave insufficient distance.' Allowance also had to be made for tides or currents using the navigational triangle, and for leeway which was pushing the ship to one side or the other.

Leaving Liverpool in September 1846, Captain Hosken broke at least two of the cardinal rules of navigation. He had failed to keep his charts up to date, to such an extent that he was unaware that a new lighthouse had been set up at St John's Bay in Ireland. And he did not believe the evidence of his dead reckoning. 'How is it possible she has not run her distance; what can have held her back?' As a result, he took the light for one of the Calf of Man lights, and supposed the other one to be hidden by the thick weather. The ship went aground in Dundrum Bay.

Dead reckoning could easily go wrong over longer distances, perhaps with a cumulative error that could put the ship many miles out of position. In 1870 she was under cloud for more than a week, and apparently recorded disappointing runs of less than 200 miles per day. Eventually it was possible to check the position and the run for that day was logged as 266 miles, but as John D. Parr observed,

I think the considerable increase of today is attributable to an unavoidable mistake of the Captain. It appears that for this last 9 days He has been obliged to make our Longitudinal distances by dead reckoning not being able to take the Sun at proper times So that to day by taking the Sun he has found a considerable eficiency [sic – deficiency?] and has accordingly added the required amount to todays log.

When sailing along or close to a coast, the navigator had to take constant compass bearings on objects on shore and plot them on the chart to make sure that he was clear of obstacles such as rocks. This became more difficult in fog and again dead reckoning came into play – a good navigator would keep it up anyway in case of sudden loss of visibility.

Ocean navigation

The first task when crossing the ocean was to choose the route. The shortest one was the great circle track. The ship would follow a curved path along a circle whose centre was also the centre of the earth. But this was often difficult; it was more important to avoid hazards and find favourable winds and currents.

The decision also depended on whether the ship was considered a steamship assisted by sail, or a sailing ship assisted by steam. The standard sailing ship route across the Atlantic involved going well south as far as Madeira then gradually turning west to pick up the trade winds. The *Great Britain* was regarded as a steamship during her Atlantic days and followed a very different and much shorter route, heading out of Liverpool and round the northern Irish coast, then directly for Newfoundland and along the coast to New York. It was often against the wind, and had many dangers. During the first commercial voyage in 1845 Captain Hosken made a fast passage of 15 days but the ship kept up her speed of 9 knots or more, through the fogs of the Newfoundland Banks and

was fortunate to avoid the numerous fishing boats and merchant ships in the area. She passed dangerously close to Sable Island, the graveyard of many a ship. Hosken was less lucky on the way back: the ship went aground off Nantucket, damaging her propeller and making it necessary to check her hull in dry dock back at Liverpool.

On the first Australian voyage, Captain Matthews ignored sailing ship practice and steamed down the African coast towards Cape Town, against the prevailing winds and the Benguela Current. She had to turn back to refuel at St Helena. The rest of the voyage was no more successful and the ship took 81 days to reach Melbourne.

For later voyages she followed the classic sailing ship route, crossing the Atlantic almost to the coast of Brazil to get the shortest path through the almost windless Doldrums – though the *Great Britain* could use her engines to reduce the time there. Rachel Henning was pleased, '. . . we were fortunate enough to steam through it, while the unhappy sailing-ships we passed were flapping their sails helplessly in the calms on the Line, while their inhabitants must have been nearly roasted alive'.

On later voyages she sailed well south and was barely within 600 miles of Cape Town, for there was no need to refuel now that engines were used as an auxiliary. At around 50 degrees south she could pick up the westerly winds that would carry her all the way, gradually turning north to make Australia. But in 1853 the

passengers, including the Henning sisters, were unhappy about the cold:

Everyone on board is very dissatisfied with coming so far South for the sake of saving four days and they all note this 'circular sailing' a failure, saying . . . that for the little advantage in point of time it gained it is hardly worth while to risk the lives of all the passengers, besides bringing us down into these cold latitudes. . . .

On the way back, it was best to use the trades again and go east, though that meant going further south to the 'howling fifties' and rounding Cape Horn at the southern tip of the Americas – 'the bugbear of passengers'. In 1862 some of them believed they were going back by the slower but more pleasant Cape of Good Hope route, but were disabused as soon as the ship was out of Melbourne.

Almost inevitably, Mr Dearlove was displeased, 'Just heard it confirmed that we are going round Cape Horn, many passengers very much disappointed & dissatisfied. They consider it a breach of faith and that they have been regularly humbugged.' But passengers were often pleasantly surprised, as the ship's newspaper reported in 1865:

During last night we rounded Cape Horn; at this famous point we dreaded fearful hurricanes, excessive colds, and any amount of icebergs, but fortunately, in all those

RIGHT The ship's binnacle, situated in front of the wheel and intended to protect the compass from the weather, while making it visible to the helmsman. *(Jonathan Falconer)*

we have been disappointed. Instead of a hurricane, there is a calm, which would make our noble vessel lie like a log in the water were it not for our 500 horse power machine.

The compass

The most vital navigational instrument, both in harbour and at sea, was the magnetic compass. Its card was divided into points. The four cardinal points of north, south, east and west were subdivided by north-east, south-west and so on. There were three more points between north and north-east, for example – north by east, north-north-east and north east by north, to give a total of 32 round the whole dial. For intricate navigation such as going into Liverpool these might be further subdivided into quarter points, such as South-East by East ¾ East.

The compass needle was subject to variation – the difference between true and magnetic north, which varied throughout the world – but charts were available for any competent navigator to correct for it. Since the early years of the century seamen had known that iron aboard the ship could also affect the compass, and of course this problem was much greater if the ship itself was made of iron – indeed some thought it might be fatal to the ocean-going iron ship.

The Astronomer Royal, Professor George Airy, conducted experiments in the ship *Rainbow* in 1838 and found that deviation could be corrected by using magnets or soft iron placed near the compass. Claxton was happy with the results and wrote to the computer pioneer Charles Babbage, 'I made a voyage in the *Rainbow*, took the bearings and distance, worked the course and perfectly satisfied myself of the perfect adjustment of Professor Airy's Apparatus. . . . I would have no hesitation in crossing any sea or latitude with his correctors.'

A ship had to be 'swung for deviation', as the *Great Britain* was after she came out of dock in 1857. The *Nautical Magazine* reported: 'This floating castle was swung to every point of the compass, and the bearings of the compasses attained with the greatest nicety.'

There were two problems that had not been taken account of. The heeling of the ship under

sail might increase deviation drastically and it was not fully appreciated until the Liverpool Compass Committee carried out some tests in 1855–60. The *Great Britain* was then refitting in Sandon Dock and the owners allowed her to be turned round to help the experiments. It was found that, 'This extraordinary ship', after much service, still had lines of no deviation which were 'much the same as . . . when she was on the stocks'.

More seriously, a ship heading south across the equator would find her deviation very much altered. In 1852 the *Great Britain* carried three compasses: one on the usual position on the binnacle, just ahead of the steering wheel, one on the skylight and one further forward, differing from one another by up to 14 degrees. By observations on the stars it was concluded that the one on the skylight was the most accurate.

Celestial navigation

An ocean navigator had to rely on astronomy, aided by his nautical almanac, which gave the position of the sun, moon, planets and selected stars at every moment during the current year. He also needed a good sextant to measure the angle of the body from the horizon, and a chronometer, a very accurate clock which compensated for changes in temperature and the movements of the ship.

It had always been easy enough to find the ship's latitude when the sky was clear. The angle of the sun at noon could be found with the sextant and then added to or subtracted from the declination – the sun's angle above or below the equator, which could be found from the almanac. The same procedure could be applied with the Pole Star in the northern hemisphere, and the Southern Cross in the south.

Longitude was much more difficult and needed some measure of time, for which a chronometer was essential. The simplest method was to observe the local time of noon from the sun and compare that with Greenwich Mean Time on the chronometer, but it was difficult to be precise about when the sun was at the top of its trajectory. The most common method was to take observations on several stars, usually near sunrise or sunset when the horizon could also be seen, or the sun some hours before or after noon. The position of each body above the earth at that moment could be found from the almanac, and a sextant reading would set the ship on a circle a fixed distance from that. Two or more readings could establish the ship's position, where they crossed. In good conditions and with skilled and diligent

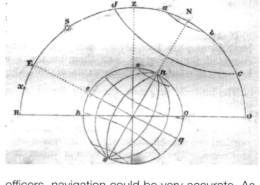

LEFT The method of taking a sextant angle. *(Author)*

officers, navigation could be very accurate. As one anonymous diarist wrote in 1871:

The excitement as we neared land, after a deprivation of its view for about 45 days, was shown by passengers remaining up until two or three of the morning as we neared Cape Otway, our first point of view was the S.E. Coast of Australia. True to time, the lights appeared, as indicated by the Ship's Officers, showing how nice the calculations of our mariners have become, by the use of improved instruments and correct calculation.

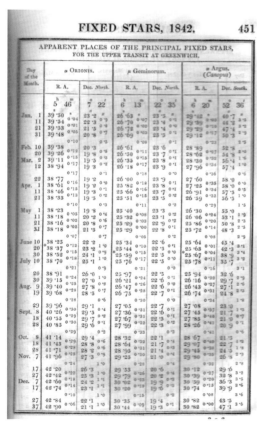

FAR LEFT A page from the 1842 *Nautical Almanac,* showing one of the main tables on the positions of heavenly bodies. *(Author)*

LEFT A star chart from Rosser's *Navigation,* used to identify the constellations. *(Author)*

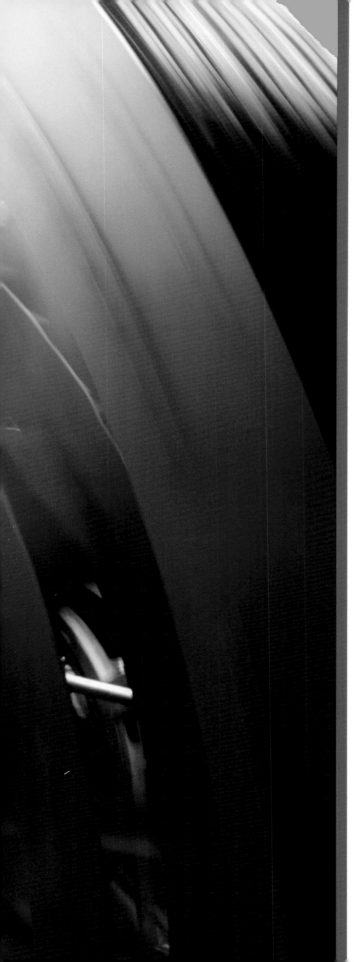

Chapter Five

The engineer's view

'Marine engines require more intelligence for their management than any other description of engine, partly in consequence of their great size and complication, and partly on account of the serious consequences an accident may entail, if the vessel be at the time in a critical situation from stormy weather or adverse winds.'

John Bourne, 1847

OPPOSITE The authentic replica of the 1,000hp steam engine gives an impression of the great power at the heart of the *Great Britain*. *(Jonathan Falconer)*

Wanted: high-calibre engineers

AMALGAMATED SOCIETY OF ENGINEERS, MACHINISTS, MILLWRIGHTS, SMITHS, AND PATTERN MAKERS.

With the rapid development of marine engineering over the previous decades, there was no pool of experienced men from which to draw ship's engineers. In 1839, after a series of accidents, a parliamentary select committee was told that steam navigation had '. . . advanced more rapidly than men of experience and knowledge can be found to conduct it; hence, we often find, in the river packets in particular, men advanced to the post of engineer who are mere automatons, ignorant of the first principles of the machinery over which they preside . . .'.

A ship like the *Great Britain*, which might be at sea for several weeks at a time, needed a higher calibre of man and it was agreed that '. . . in our first-class steamers, the engineers have from the start been hand-picked men'.

Perhaps they were among those of Robert Spencer Robinson's acquaintance: 'The engineers whom I have personally known in merchant steamers, were men of a most excellent description, well-educated, sober, steady, and with a perfect knowledge of their business.'

Captain Edward Blackmore, looking back on the period, believed that most engineers

. . . had not served an apprenticeship in the trade but were engine drivers raised from the shovel – many were good, steady fellows who handled well the slow-going machinery of the times, with its low pressures – most of them a good deal better than a chance engineer out of a shop put in charge of machinery afloat with little or no experience as a sea-going engineer.

Engineers and firemen

Civil and mechanical engineers like James Watt, Thomas Telford, George Stephenson and Brunel himself were national heroes, but those who operated ship's engines were self-effacing. They were rarely noticed by passengers except in May 1860 when the saloon passengers informed Chief Engineer Noah Beamish that 'we are all fully sensible to you in the successful accomplishment of our voyage'.

There are very few biographies or memoirs to tell us about their habits and thoughts. The profession had originated in the trade of millwright, men who knew something about machinery and metal-working. Some had experience on static steam engines in factories, or in locomotives – though Brunel regarded the latter as an inferior breed. Some were trained from among the seamen, which ship-owners and masters preferred, for 'the reason that mechanics would frequently not endure them, being liable to be treated as seamen by the captains'.

On the first Atlantic voyage of 1845, the engine-room department was headed by Henry Clements with six skilled engineers under him. There were 18 firemen to stoke the boilers, presumably better than the crew which had brought the ship round to Liverpool – 'a set of oblivious, idle and independent firemen'. The firemen were assisted by ten coal trimmers, to get the coal out of the bunkers ready for use.

In 1847 John Bourne suggested, 'In dealing with the firemen and coal trimmers, there must be as many standing orders as possible, so that the engineer may be saved a perpetual expenditure of thought on matters of routine, where by his attention may be given to more important objects.' Tidiness was essential. 'For everything in the engine room there should be a place, and for its being there, and being clean and fit for use, some person or other should be responsible.' But firemen continued to be unreliable, and in 1852, the trip to New York was almost delayed because some of the firemen were drunk. Rosamond D'Ouseley was rather shocked by her visit to the boiler room in 1869: 'We also went into the Furnace Rooms & were nearly roasted, the men who work them seldom live to be more than forty.'

While the seaman was out in the open in all weathers, the engineer was below decks in an environment which was nearly always far too hot for comfort. The seaman had to work hard when raising anchor or taking in sails in an emergency, but for most of his time on watch he was standing by, perhaps cleaning decks or carrying out maintenance work at a gentle pace. He heard the sound of the wind and waves and the creak of the masts, but otherwise it was quiet, in contrast to the engine room. The fireman had to work hard and constantly when the engines were running, with the heavy labour of shovelling coal. It was generally recognised that a good seaman had to start at an early age, whereas a fireman needed to be strong and therefore fully grown. Captains often tried to get seamen to help in the engine room but that was rarely successful. According to the traditional sailor's phrase, 'oil and water don't mix'.

The steam engine

Like its successor the internal combustion, the steam engine of the day used pistons inside cylinders and had a crankshaft to transmit the motion in the form it was needed. But in another sense it operated on the opposite principle – the contraction or condensation of a gas (steam) rather than its expansion. When condensed into water, steam takes up one 1,600th of its previous volume. This created a vacuum which sucked the piston in one direction.

James Watt's greatest idea, nearly 80 years before the *Great Britain*, was to pump the steam out into a separate condenser so that energy was not wasted with alternate heating and cooling inside the cylinder. Watt also developed the double-acting engine, so that as the steam was withdrawn from one side of the piston in each cylinder, more would enter from the other side. The *Great Britain*'s engines largely followed the principles of Watt, including the low steam pressure of 5lb above atmospheric.

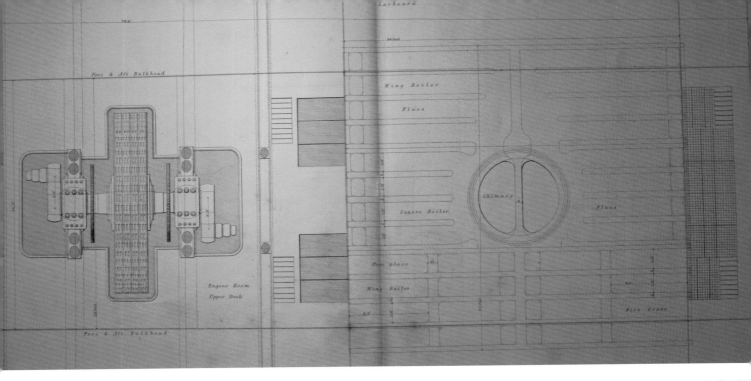

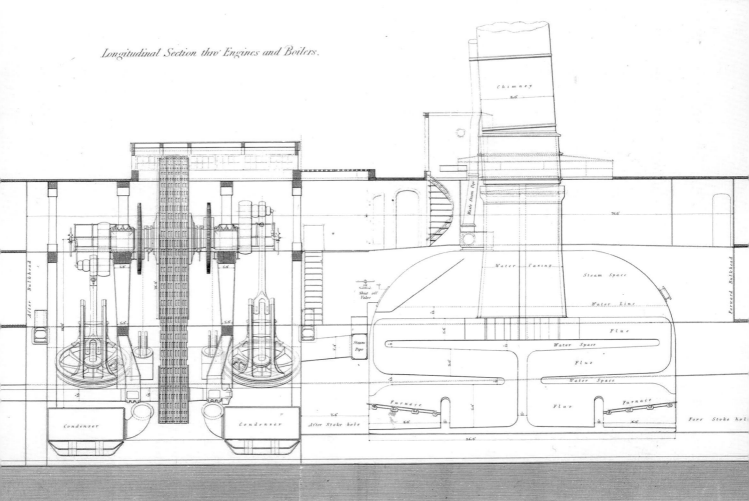

Longitudinal Section thro' Engines and Boilers.

Fuel

Coal was the essential fuel of the ship. It was loaded on board by the crew or by hired labour, for example by 'coolies' when refuelling at Cape Town in 1852. It was stored in bunkers as close as possible to the engine, not just for convenience but so that the trim of the ship would not be altered by sudden use of coal near the bows or stern. Fuel could also be stored on deck, but that created instability, and made the ship very dirty to the distress of the passengers.

Coal came in different varieties, which was important on a long voyage. 'Best Welsh' had a calorific value of 9.5lb of water evaporated per pound of coal, 'average Welsh' of 8.04 and 'average large Newcastle' of 7.66. Anthracite had the high value of 9.01, but it was not easy to ignite or break up, and it would not adhere to the sides of a furnace, which could be a problem in rough seas.

During the trial voyage in 1852, two tons of Welsh coal were used for getting up steam, followed by 26 tons of anthracite. During the subsequent voyage to Australia, Captain Matthews experimented with a mixture of anthracite and patent fuel. Since he hoped to make most of the voyage to Cape Town with little wind assistance, the decks were stacked with bags of coal, which inconvenienced the

passengers and exhausted the seamen who had to move them. Despite Matthews' vigilance, he was shocked to find that the record of coal used per day did not tally with the amount left on board – there were only 250 rather than 450 tons, and the ship had to turn back to St Helena. He acquired both coal and wood there, and used them in a mixture, but it was difficult to get enough steam.

Boilers

In 1852 Robert Murray wrote,

> *The management of the fires on board a steam vessel affects the question of economy in the consumption of coal to so great an extent, that the importance of skilful firemen cannot be too much insisted upon. It is a great mistake to suppose, as too many captains and owners of steam vessels do, that any able-bodied man who can throw coals on a fire is fit for a stoker.*

The only exception was in the case of auxiliary screw ships, which 'may be expected to make a considerable portion of their voyage under canvas' – rather like the *Great Britain* in her Australian years. Even then, there should be a well-trained fireman for each watch to supervise the seamen acting as stokers. Otherwise,

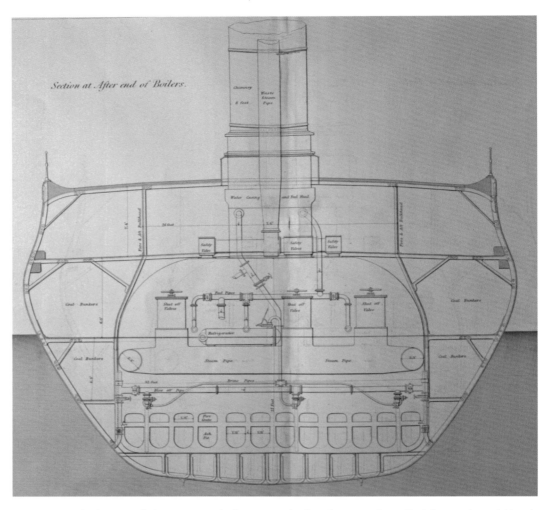

Section at After end of Boilers.

*. . . many instances of steamers are to be
found at the present day in which the same
quantity of coal is regularly burnt per hour,
whether the engines are going fast or slow
. . . in such a case, one quarter of the fuel
is thrown away through ignorance of the
despised art of stoking.*

Skilled stokers would keep the fire burning
at the right level of brightness, level off the coal
above the bars, shut the fire doors as quickly as
possible, prevent the coal from caking, re-burn
half-consumed cinders and leave the fires clean
at the end of a watch.

As fitted in 1843, the *Great Britain* used
the 'flue' type of boiler, rather than the newer
tubular boiler. Each was 34ft long, 31ft wide
and 21ft 8in high with a domed top. They were
divided into six separate boilers, three forward
and three aft, each fed with four furnaces. Each
of these contained a grate made up of bars

sloping downwards so that the coal would tend
to move towards the brick-built 'bridge' near
the centre of the boiler. The fireman fed coal
in through the furnace door, and the gratings
allowed cinders and ash to fall into the ash pit
from where they could be cleared. The heat
from the furnaces passed through a series
of flues inside the water tank, which were
arranged to expose 1,248sq ft directly to the
fire, plus nearly 10,000sq ft more by the flues.
After doing its work, the smoke and heat from
the separate furnaces exited through the funnel.

The watchkeeper had to keep a good eye on
the glass water gauge attached to each boiler.
The level had to be high enough or parts of the
metal would overheat and cause an explosion;
but if it was too high, it might boil over and get
into the steam pipe and cylinders, causing a
breakdown. There was a danger that too much
salt accumulated within, clogging up the pipes and
preventing the heat reaching the water, so every

two hours or so the stoker had to 'blow off' the highly salinated water which might accumulate in the lower part of the boilers. By 1852, they were fitted with a salinometer to measure the density. Each boiler had a safety valve which would let off steam if the pressure became too high.

The boilers needed a good deal of maintenance, both afloat and in harbour. When steaming, it was common to keep two or more of them out of service so that the fires could be cleaned. Deeper maintenance had to be done in harbour, as described by Robert Spencer Robinson:

> On the vessel's anchoring, and the steam being no longer required, the fires are drawn and extinguished, and the water in the boilers driven violently out of the blow off pipes by the expansive force of the steam. . . . Of course if the boiler has to be cleaned out, which should be done whenever there is time enough for it, the man hole door is taken off, and plenty of air necessarily admitted.
>
> The parts of iron boilers most likely to decay, when proper attention is paid to the interior, are the lower parts of the shell, in situations where they cannot be seen, or got at, wherever, from any cause, salt water gets at them, and is alternately wetting the shell and then leaving the salt to form incrustations, &c.

Engines

Once sufficient steam had been worked up in the boilers, the engineers took care in starting the engines. It was important that no water or air should remain in the cylinders. Water was incompressible and 'if the piston should come into contact with it at either end of the cylinder, and there be no escape for it, something must break, a sway beam, a side rod, or a piston itself'. To prevent this, the engine was turned by hand through several revolutions using wheels on the upper deck. The engine would not start if the pistons were near the end of their movements, so the turning was stopped with the crankshaft horizontal.

The engineer controlled the rate of the entry of steam by means of the throttle and expansion valves. According to Robert Murray, the engineering writer, 'The throttle valve of a marine

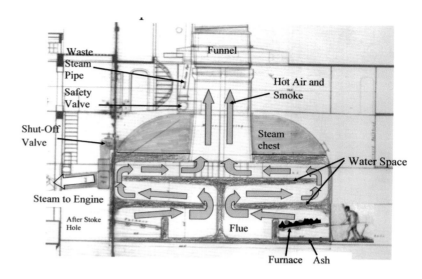

engine is always worked by hand, and should be used only in controlling the speed of the engine for any temporary purpose, such as passing through a crowded river, before stopping at a pier, &c., but should seldom or never be used for working the engines expansively at a permanent reduction of speed.' That was to be done by means of the expansion valve.

In the *Great Britain*, the throttle was a butterfly valve at the end of the steam pipe just outside the slide valve. Next to it was the expansion valve, which was a series of eight rectangular holes in two rows. The engine could

ABOVE The operation of the boilers showing the paths of the smoke and steam. *(Author)*

BELOW A cutaway drawing showing the steam engine installation in the *Great Britain*. *(Illustrated London News)*

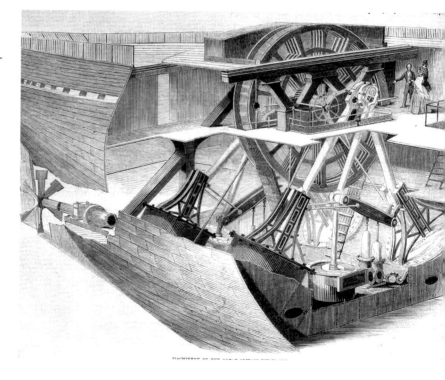

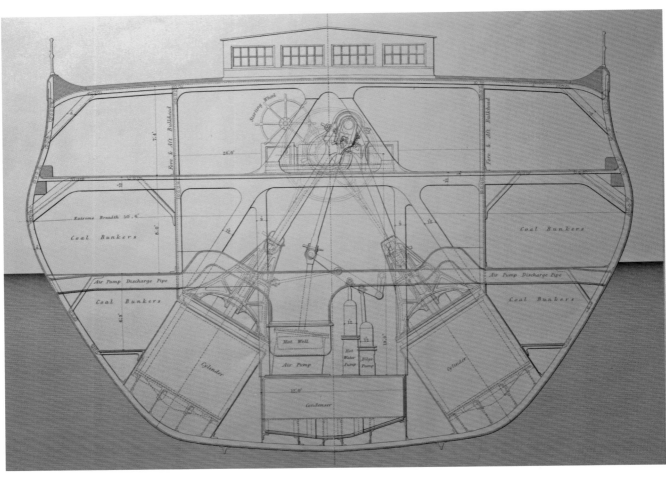

Labels on the cross-section diagram:
- Steering Wheel
- Fore & Aft Bulkhead
- 26'.6"
- Fore & Aft Bulkhead
- Extreme Breadth 50'.6"
- Coal Bunkers
- Coal Bunkers
- Air Pump Discharge Pipe
- Air Pump Discharge Pipe
- Coal Bunkers
- Coal Bunkers
- Hot Well
- Air Pump
- Hot Water Pump
- Bilge Pump
- Cylinder
- Cylinder
- 15'.0"
- Condenser

ABOVE This cross-section through the engine shows the pistons and the great wheel, which turned the lower shaft and propellers, and the hot well and condenser. *(ss GB Trust)*

RIGHT Reversing levers to control the expansion gear. *(Jonathan Falconer)*

BELOW Part of the gearing of the replica engine. *(Jonathan Falconer)*

reach maximum power if steam continued to be let in during the whole of the stroke, but this was inefficient and it was agreed that it should be gradually cut off as the stroke progressed. It was found that the *Great Britain*'s engines were most efficient at a rate of 6:1, that is the steam was cut off after the first 12in of the 72in stroke.

The steam went into the slide valve, which was set to allow it alternately into the top or bottom side of the piston, while letting it exit to the condenser from the other half. The *Great Britain* used a slightly unusual but very well-designed type of valve with a small piston moving up and down through 4½in. The exit ports from the slide valve to the different parts of the cylinder were angled to save wear and tear and the system was carefully balanced. The piston was controlled by the eccentric gear which was linked to the crankshaft. This consisted of an 8ft wheel with a rod set off its centre, or eccentrically. The engine was put into reverse by changing the sequence of the steam input to the cylinder rather than by gearing. The more sophisticated Stephenson link was not available and it was necessary to disconnect the rod from the wheel and set it in another hole on the opposite side of the wheel.

The four cylinders were 88in in internal diameter and with a stroke, or movement of the piston, of 72in. The cylinder heads and feet were conical, pointing downwards. Inside each was a piston which replicated the conical form and was fitted with split piston rings, rather like a modern car engine, to give a tight fit. The engineer used the indicator diagram to monitor the pressure inside the cylinder. This was recorded by a pencil which moved up and down with the pressure, and marked a sheet of paper which revolved with the movement of the cylinder.

For each revolution, the steam entered through the slide valve and the pencil rose to the 'starting point' at the top of the diagram, probably at a pressure of 5lb above atmospheric. The line turned a fairly sharp corner as the cylinder became full, then began to move slowly downwards as the supply of steam was shut off during the expansion phase, after which the pencil marked the 'expansion corner'.

At the end of the stroke the steam was sucked out and the pressure declined rapidly to perhaps 10lb below atmospheric, a partial vacuum, at the 'eduction corner'. This pressure remained fairly constant during the expansion of the steam on the other side of the piston. When it reached the end of the stroke at the 'lead corner' of the diagram, steam was let in so that pressure rose sharply and the cycle began again.

The indicator diagram could show up any faults of the engine. Ideally, each stroke would be drawn on top of the last one – if not, the engine was operating irregularly. In addition, according to John Bourne,

BELOW A typical indicator diagram. *(Author)*

BOTTOM The actual indicator diagrams from Brunel's screw ship *Rattler,* as used in the famous trial against the paddle steamer *Alecto* in 1845. *(Author)*

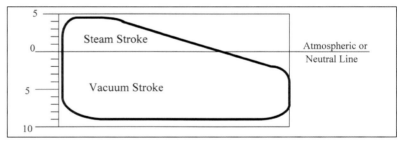

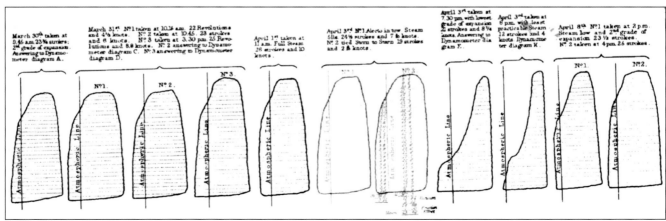

If the 'eduction corner' be much rounded or slanted away, then the eduction passages are too small. If the 'lead corner' be much slanted off, then the amount of lead given to the engine is great, or in other words, the steam side of the slide is opened before the end of the stroke. If the 'starting corner', or the steam corner, as it might be called, is slanted off, then the steam does not gain admission to the engine sufficiently early.

Conventionally the power of an engine was measured by a formula based on the volume of the cylinders and the length of the stroke – the *Great Britain*'s were measured at 1,000 nominal horse power by these means. But this took no account of the real efficiency of the engine, improvements in technology since the formula was devised by James Watt, or losses through friction, leakages and so on.

The *Nautical Magazine* reported in 1843 in connection with the *Great Britain*, '. . . nominal horse-power is a most inaccurate basis for calculations' and that 'no reliance can be placed on estimates deduced from such data'. The indicator diagram could be used to calculate a much more realistic figure. The area enclosed by it was measured and from that the mean pressure per square inch on the cylinder could be calculated. This was multiplied by the area and speed of the piston then divided by 33,000 to give the notional work of a horse in a day. The result was the indicated horse power, which was often two or three times the nominal horse power – for the *Great Britain* it was 1,800, which was rather low in proportion to nominal horse power.

Meanwhile, the steam was sucked out of the cylinders through the eduction pipes and went into the condensers. There were two of these, each serving a pair of cylinders, and fitted with an injection pipe which brought in cold seawater.

. . . from this pipe a shower of water constantly plays into the condenser while the engine is working; the water is injected in such a manner that the steam entering the cylinder must mix with it, and is instantly condensed, by which means a vacuum is made in that end of the cylinder which

is for the time in communication with the condenser, thus taking away all obstruction from one side of the piston, and giving no hindrance to the pressure of steam upon the other side.

The water created in the condenser was taken out by air pumps, which were also driven by the crankshaft. It went into a container known as the hot well until it was pumped back to the boilers to begin the process again. The engine was also used to drive the bilge pumps, which relieved the seaman of the arduous task of pumping water out of the hold.

While this was going on, the motion created within the cylinder was transmitted diagonally upwards by the piston rods which extended some way beyond the end of the cylinder, guided by roughly triangular castings known as the cross-heads, which were an alternative to the heavy beam used on many engines, or the more complex parallel motion system. The ends of the piston rods were pivoted on the connecting rods. This began the process of translating the reciprocal motion produced by the cylinder into the rotary motion needed for the propeller. The upper end of the connecting rod was pivoted on the crankshaft by means of the crank pin, a crucial component which was subject to great stress and strain and was difficult to cool and lubricate. The crankshaft was a round wrought-iron bar with a maximum diameter of 2ft 4in. L-shaped pieces were attached to both ends, for the connecting rods of each pair of cylinders.

The engineer had to check constantly that bearings were not too loose, which would cause vibration, or too tight, which would make them overheat. Lubrication was mostly done by animal and vegetable products such as tallow and linseed oil. Some bearings could be lubricated by means of oil cups placed above them and dripping oil.

According to Bourne, 'It is usual to lubricate the crank pin, when it heats, with sulphur and tallow, which is supplied through a funnel cup instead of the ordinary oil cup. . . .' Another crucial area was the thrust bearing at the head of the propeller shaft, which had to bear all the force that drove the ship forward under steam – up to 20 tons when accelerating.

This was cooled by pumping water in under pressure.

Stopping the engine needed some care, according to Robinson:

The engine should always, except in case of absolute necessity, be slowed or eased, before it is stopped. When intending to stop, the throttle valve and injection being shut off, the engineer watches the eccentric rod, closing the ports, and immediately throws it out of gear, ships the starting bar ready for a back turn. If the wheels have great velocity at the moment of stopping, they will perhaps revolve a turn or two from their momentum, compressing the steam in the cylinder, and incurring much risk to the machinery.

Driving the propeller

The engines rotated at about a third of the speed needed for efficient operation of the screw. Brunel rejected the idea of gearing, which tended to be noisy. Instead he adopted a system using chains, whose manufacture was described by Guppy:

These links were very carefully forged, they were then brought to a dull red heat and placed in a proving machine, where they were stretched one-eighth of an inch, and while in that state they were rigidly examined. After boring and planning, they were all finished on one gauging and case-hardened.

The propeller shaft was made in three sections. The innermost one passed through a watertight door. The intermediate one, 61ft 8in long and 30in internal diameter, was hollow, made up of curved plates riveted together with countersunk heads. Aft of that was the tailshaft, 25ft 6in long. The inner end of the shaft was mounted on a thrust block, as described by Guppy:

The thrust, or effort of the screw, is received by a step, composed of a steel plate 2 feet in diameter, against which a gun-metal plate, of similar diameter, affixed to the head of the shaft, presses. A stream of water is admitted to a cavity, in the centre of these plates, and very satisfactorily lubricates them.

Propellers

The first propeller fitted to the *Great Britain*, and the one which is replicated on the ship today, was described by Guppy:

The screw of the 'Great Britain', which is of wrought iron, consist of six arms, formed by placing and riveting together four distinct forgings, or centre pieces, with arms welded to them, each of 6 inches thick.

Upon the extremities of these are riveted

ABOVE Two giant wheels and a set of chains linked the drive shaft to the propeller shaft running along the bottom of the ship. *(Jonathan Falconer)*

LEFT The replica of the original propeller and the balanced rudder. *(Jonathan Falconer)*

*palms of plate iron which are 4 feet 4½
inches long on their circumferential edge, by
2 feet 9 inches in height, and ⅞ inch thick.*

Despite its rather crude appearance, it was
remarkably efficient. Guppy wrote, 'the

present screw has rather surpassed my
expectations . . .'. In more recent times it
was tested at the Vickers Experimental Tank
and proved almost as efficient as a modern
propeller. However, it was not strong enough
to withstand the accidents which befell it under
Captain James Hosken, when all its blades
were lost and the ship had to complete the
trip back to Liverpool under sail. Fortunately a
replacement was already in hand.

A very different four-bladed propeller had
been designed by Brunel and cast after the
original trials. It had the same diameter and
pitch, but was heavier at 7 tons, and more
robust, though probably less efficient. The
propeller could be disengaged for sailing, but
the engine had to be stopped and four screw
bolts taken out before drawing back a coupling
box. This took about a quarter of an hour, and
some considered the process too complex.

Communication with the engine room was
by speaking tube in the early days, and later
by a very early engine room telegraph, as
described by Mr Dearlove: 'Communication is
held between the deck and Engine room by
means of a Telegraph speaking tube & Bell. The
Bell calls first the engineers attention and then
as the hands on the dial on deck are moved to
ahead, Stern & C: so the hands on the dial in
the Engine Room do the same.'

New engines

In 1852 the ship was fitted with new and more
advanced engines, designed and built by the
firm of John Penn of Greenwich, which was
already the main supplier to the Royal Navy. It
took the lead in developing the oscillating engine,
in which the cylinders themselves swivelled to
eliminate the need for gear such as cross-heads.
According to Robert Murray in 1852:

. . . the oscillating engine, which has derived from Mr. Penn so much of the elegant simplicity and present perfection of workmanship and arrangement, is generally preferred. It need hardly be explained that this engine derives its name from the fact of the cylinders 'oscillating' upon hollow axes or 'trunnions', through which the steam is admitted to, and withdrawn from, the valves – the piston rod by this means accommodating itself to the motion of the crank without any 'parallel motion' being required.

It was claimed by John Bourne, 'the performance of the "Great Britain" with these engines is superior to her previous performance with engines of twice the nominal power. Without the aid of sails and with a full cargo she realises a speed of fully 10 knots per hour.'

In the first Australian voyage of 1852, when the engines were in almost constant use, Chief Engineer Reginald Bright kept a log with columns describing the hourly expenditure of coal in hundredweights, the readings on the steam and vacuum gauges, the revolutions of the engine, the numbers of boilers at work, the

LEFT A speaking tube in the engine room was used to convey orders from the bridge. *(Jonathan Falconer)*

type of coal used and the engineer on duty. He also recorded the rate of the ship's log, the courses steered, the distance run, latitude and longitude at the end of the period and the revolutions per day and the wind and weather. This gave the basis to assess the efficiency of the engines in different conditions by calculating how much coal was used and revolutions turned to make a certain amount of progress. He kept daily accounts of the consumption of tallow, engine oil, linseed oil and cotton waste for cleaning, and described the maintenance work done. The screw recorded around 25,000 revolutions in a typical day's work.

The new engine seems to have caused more vibration and one woman complained: '. . . She used to ask the captain in the morning, after the

BELOW Cross-sections of the oscillating engines and tubular boilers fitted in 1852. *(ss GB Trust)*

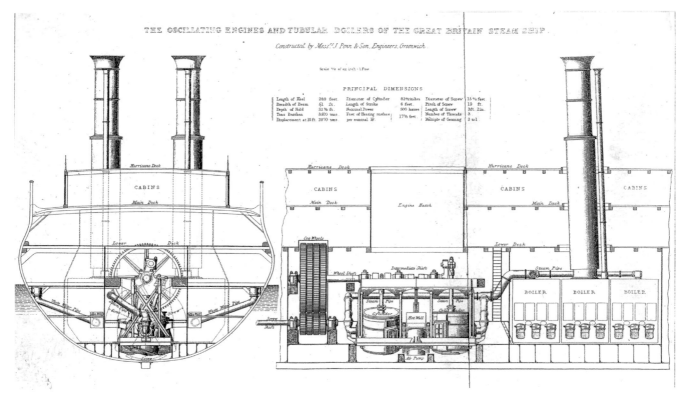

THE OSCILLATING ENGINES AND TUBULAR BOILERS OF THE GREAT BRITAIN STEAM SHIP.

Constructed by Mess.rs J. Penn & Son, Engineers, Greenwich.

PRINCIPAL DIMENSIONS

screw had been shaking the ship a good deal, if they had not been going over some rocks because she was sure they had as she had "felt the ship wriggling over them". . . .'

The boilers were on the more advanced tubular principle, as described in 1862: 'The boilers are situated forward of the Engines which is called a Stoke Hole, the Boilers are all separated from each other. They are on the tubular principle, each containing we are told 280 Brass tubes, they are each about 11 feet long and 13 feet deep.' They were situated with the fire doors facing towards the sides of the ship, which may have made it more difficult for the stokers in bad weather.

New propellers

The ship was fitted with her third propeller at the same time as the new engine. On this occasion it was three-bladed, a casting of the same diameter as the other two but with less pitch. It could be disconnected in 10 minutes according to the log of 1852, but it caused serious problems when Captain Mathews tried it at seven in the morning of 27 September 1852. According to a steerage passenger, 'the ship came almost to a standstill'. John Gray distinguished himself in rallying the crew with his 'gift of ubiquity', but the fore and topmast studding sail booms broke before he could do anything. The steerage passengers 'were all requested to assist in taking in sail'. The ship had to be stopped for the screw to be reconnected.

Perhaps this is why the fourth propeller was fitted during the next major refit in 1857. Based on naval experience, it was two-bladed so that it could be lifted into the hull and a special frame was cast for it to operate in. Mr Dearlove described the technique in 1862:

When the screw is unshipped, the shaft is drawn inwards telescope fashion for about 5 feet it requires an engineer and five men to effect this purpose in the alley whilst on deck 20 or 30 men are at the after capstan round which a strong chain is suspended over the Chamber abaft the wheel up which the screw is drawn up or lowered as required and properly secured.

Maintenance of the hull

The hull of the *Great Britain* was far more durable than any wooden ship, and it was very well built, so a great deal of it survived intact through the decades – in contrast with wooden ships such as HMS *Victory*, where only a small proportion of the original timber survives. The *Great Britain* carried a blacksmith, mostly for repairs on the engines rather than the hull. It was noticed in 1862, 'There is also a Blacksmith shop over one of the Boilers where you may see a Blacksmith busy at work as there is always plenty of work in this floating village.'

In addition, the ship carried boilermakers, and most of the engineers were skilled in light metal work. The ship did not carry divers (but hired one to clear the propeller at Malta in 1855), so running repairs to the hull could only be done from the inside, by strengthening supports and so on. This was first carried out after the storm during her trials, when the forecastle was damaged and light wrought-iron stanchions were apparently fitted in the area and are extant today. After the second voyage to New York, workmen in the city were employed to help the crew to repair many broken angle irons and butt plates in the bottom, as well as replace the worn chains of the engine.

More extensive work needed a dry dock. After the grounding in Dundrum Bay, 150ft of plating had to be replaced while deeper keelsons were fitted inside, and the wooden stringers under the deck were augmented or replaced with iron. During the 1856–57 refit, it was reported that 'a great number of workmen were continuously employed, and the ship was subject daily to small concussions and vibrations'. By that time this 'extraordinary ship' had been 'stranded, and altered, has traversed both hemispheres, and been very many years in active service' – but her structure was largely intact. In 1872 she was surveyed by Lloyd's and it was found that the plates under the engines had been renewed for a second time in 1861. The thickness of her plates generally had reduced by 15 or 20 per cent through the normal action of the sea. But despite changing standards since she was built, the plating was 'well wrought' and the riveting was 'good'.

Repairs to the engines

Both sets of engines used by the *Great Britain* were quite robust, but some breakdowns were inevitable with long voyages and hard service. In 1846 the drum wheels and drive chains of the transmission were found to be heavily worn and had to be replaced hurriedly in New York. Less seriously, during the 1852 voyage to Australia there were several cases of individual cogs being reported loose and it was necessary to stop the engines to repair.

There was trouble in May 1855, just after the ship had entered the Black Sea on the way to the Crimea. The cover of one of the hot wells burst, spraying scalding water into the bilges. Chief Engineer Alexander MacLennan stopped the engines instantly and the seamen had to empty the bilges without them. It was found that the cover, supposed to be an inch thick, was very uneven and less than a quarter of an inch in places. MacLennan and his men worked for 24 hours with 'wood planks, wedges, shores &c.' to make an improvised cover so that the ship could steam at 6 knots. She anchored in Kalmiesh Bay and the engineers worked day and night to produce a ¾in iron cover that would be 'stronger than ever before'. MacLennan was proud that '. . . we repaired it ourselves without any assistance or Expense from Shore'.

But the most serious breakdown came just

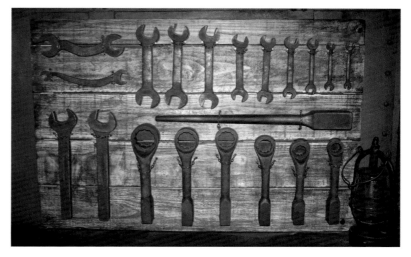

off the Australian coast near the end of the voyage. Most of the passengers were asleep or just beginning to rise on the morning of 11 February 1868 when they were disturbed by 'a tremendous noise, and the ship received such a shock as almost to make us fear she was going to the bottom'. Then the engines began to race until they were shut off by the engineers. Passengers rushed on deck, many half dressed. The ship was in danger of being driven on to Cape Otway. It was soon established that the propeller shaft had broken in the hollow section and the engineers worked feverishly to repair it. Eventually they were able to drive in four rivets, which allowed the ship to move at half speed and reach her berth off Melbourne.

ABOVE A selection of tools in the *Great Britain*'s engine room. (*Jonathan Falconer*)

LEFT A close attention to detail has been paid in the creation of the replica engine. (*Jonathan Falconer*)

Chapter Six

The passengers' view

In the Victorian age there was no cruise director or hotel manager to form a link with the passengers. The purser and surgeon were not always up to the task, so it was very important that the captain could get on with them. The SS *Great Britain*'s Captain Gray was exceptionally good at this and passengers wrote of 'the intimate relationship which has subsisted between yourself and us'.

OPPOSITE Graffiti carved into the ship's timbers by the crew.
(Jonathan Falconer)

The crew

The domestic life of the ship was nominally the responsibility of the purser, though he does not figure largely in passengers' accounts, and he was not always a strong personality. On the ninth voyage, John Anjer was described as 'an honest man but very slow'. His assistant, the storekeeper John Waddle, was 'Very slow, and drinks, it is thought. The coffee to the fore saloon was only served out half the quantity of scale. Waddle's fault.'

The chief steward in 1872, John Campbell, had already served 16 years at sea. He was much concerned with the animals carried on board, which he regarded as 'my livestock'. He was less happy with female passengers. 'I do wish that women would never come to sea. They are the top and bottom of every trouble. . . .' In 1870 the passengers offered their 'heartfelt thanks to him' for 'his kindness and civility to us during the voyage'.

Passengers got on well with the stewards. From the saloon Rosamond D'Ouseley wrote, 'We have very agreeable stewards, Mr Melville and the Head Centre, who having once been an officers servant marches about like a grenadier & is particularly attentive to the ladies.' In 1870

one man in the steerage was 'our very little stewart, he is a very little man, about as little as I ever saw. Only he has not been anyways contrary to any of us, and he was doing his best to keep things clean, which was not very easily done with such a rum lott.' Mary Compton noted their skill:

> We have 12 stewards to wait for dinner, 2 for each table, it is very funny to watch them bringing in the dishes: one rings a little bell which is the signal for the cook to send up things from the lower regions, then the 12 stand in a row down the saloon passing the dishes like firemen pass buckets; on rough days they have to stand slanting and look very comical, they very seldom drop the things; one spilt a cup of tea down a gentleman's back yesterday.

But Mr Dearlove, a constant complainer, had a different view about 'each little petty officer steward &c. drest in a little brief authority [who] arrogate to themselves as many airs as a Spanish grandee . . .'.

Stewardesses were employed in small numbers on most of the later voyages, usually with one or two on board to cater specifically to

RIGHT Crew members on the weather deck of the *Great Britain* in the 1880s. Note the different ethnicities among the crew, and the hens strolling on the deck. *(ss GB Trust)*

women's needs. According to Mother Mary Paul Mulquin, theirs was '. . . like all of her class, a busy, smart, Scotch woman with a remarkably broad twang of accent, she promised to be all of attention to us, and indeed she does not lead an idle life, having 43 beds to make up daily, besides all the other items of Ladies Toilette'.

Other members of the crew who worked with the passengers included barkeepers and bakers, though butchers were not always successful – in 1863 a notice appeared reading, 'Wanted – a good butcher to commence business near the fore saloon. . . .' There was a barber in 1861–62, though to modern eyes Mr Dearlove was patronising to say the least: 'Jones the coloured barber requested permission after tea in our cabin to read some poetry composed by himself. . . . It was really very creditable and no disgrace to any man therefore it surprised us more by been composed by a coloured man.'

The passengers

We know little about the inner life of the ship during her New York period. The voyage was comparatively short and passengers were few, so they enjoyed a large

amount of space and the attentions of a full complement of stewards. They were mostly wealthy and many of them were experienced travellers who did not regard such a voyage as remarkable – in any case, none of them are known to have kept diaries. In the fateful voyage of September 1846 it was reported that the passengers were 'principally Americans, French and a few English'.

When converted for the Australian run, the ship carried a much greater variety of passengers, in larger numbers and for a much longer voyage. W.D. Waters commented, 'We have on board some of all classes of Society from the highest to the very lowest. I have never mixed with such a lot before. Some of the lowest of the Irish people are sent by the

BELOW A poster advertising the voyage from Liverpool to Australia aboard 'the celebrated auxiliary steam-ship *Great Britain*', 25 October 1873. *(ss GB Trust)*

LIVERPOOL & AUSTRALIAN NAVIGATION CO.

STEAM FROM LIVERPOOL TO AUSTRALIA.

THE CELEBRATED AUXILIARY STEAM-SHIP

GREAT BRITAIN,

3209 Tons, and 500 Horse-power,

CHARLES CHAPMAN, Commander,
IS APPOINTED TO LEAVE THE RIVER MERSEY.

FOR MELBOURNE AND BRISBANE

(Landing Passengers and Cargo at Melbourne, and proceeding without delay);
TAKING PASSENGERS ALSO FOR

SYDNEY, ADELAIDE, AND NEW ZEALAND,

ON SATURDAY THE 25th OCTOBER, 1873.

This magnificent and far-famed Ship **has made the passage out to Melbourne in the unprecedented short time of 53 days.** She affords an opportunity for Passengers to reach Australia in almost as short a time as by the Overland Route, *viâ* Southampton, without incurring the very heavy expenses attendant thereon, and avoiding entirely the discomfort of frequent changes. Her Saloon arrangements are perfect, and combine every possible convenience, Ladies' Boudoir, Baths, etc.; and her noble passenger decks, lighted at intervals by sideports, afford unrivalled accommodation for all classes.

FARES,

Including Steward's Fees, the attendance of an experienced Surgeon, and all Provisions of the best quality.

		TO MELBOURNE.	TO BRISBANE.
AFTER SALOON {POOP		60 and 70 Guineas	66 and 76 Guineas
BELOW		55 „ 60 „	61 „ 66 „
SECOND CLASS (on Deck)		35 „ 30 „	28 „ 33 „
THIRD CLASS		18 „ 20 „	21 „ 23 „
STEERAGE		15 „ 16 „	18 „ 19 „

Children under Twelve Years, Half-price.   Infants under Twelve Months, Free.

Government into the steerage of the *Britain* to be imported into the Colony and we have got no less than 250 of this class upon board this time.'

With the building of the Suez Canal and the transcontinental railway across the USA, the *Great Britain* was no longer the fastest way to Australia by the 1870s and an anonymous passenger considered his options.

The route by New York and San Francisco, over the wide continent of America on the Union Pacific and Rio Grand Lines of railway incessantly for 7 days, and then for 30 days over the Pacific Ocean. . . . This is costly, but interesting. . . . The most direct and select route, by which our letters usually travel, is that from Southampton, Egypt, Bombay, Point de Galle to Melbourne round the S.W. side of Australia conducted by the celebrated P & O Company. . . . But by it there is inconvenience where much baggage is taken. The route I selected was from Liverpool to Melbourne, by the 'Great Britain', a ship at the time the largest passenger ship in the world, and which has made more prosperous voyages than perhaps any ship now afloat.

Emigration to Australia had changed in character by the 1860s, and it was now undertaken by 'respectable' artisans and clerks who hoped to better themselves. Some of them might have read books like *Another England* of 1869, whose title said a good deal about its message. A former official of the government of Victoria claimed that the old days of convicts and the gold rush were over. 'The country has now sown its wild oats, and has settled into a respectable mode of life; a respectability happily attended with competence and comfort.'

Many passengers specifically chose the *Great Britain*. According to Seaman Felix Greyner, 'She had a great name at that time as a Passenger Ship, and was always full both out and home.' Rachel Henning was already familiar with the ship as her sisters had sailed on her before she left from Liverpool herself in 1861.

Passengers returning from Australia were not necessarily going with their tails between their legs, possibly because failures could not afford the fares of the *Great Britain*. It was reported by Robert Caldwell, on board the ship in 1855:

We had about twenty-three passengers in the salon: three married couples returning to see friends and educate their families – all had made fortunes; nine single men, most of them understood to have incomes of from £1,000 to £5,000, and suspected of intending to get married, and all intending to return; two young ladies, one a governess; four doctors who had taken out emigrant ships; one a captain and his wife, who had taken out a ship. In the other parts of the ship several families were returning with moderate fortunes – they had mostly been working people. A large portion were gold-diggers, who had made money. . . . Very few, indeed, either admitted or had the appearance of being unsuccessful.

One returnee avoided the *Great Britain* because she was so popular. 'The *Great Britain* will monopolize the bulk of the passengers so that we shall leave with about 50 or so in all

ABOVE LEFT A passenger ticket for nine members of a family to go to Australia on the *Great Britain* in 1862. *(ss GB Trust)*

ABOVE A second-class passengers' contract ticket for a family of four to sail to Melbourne, Australia, in 1867. The full ticket cost of £78 15s has the equivalent purchasing power today of about £5,440. *(ss GB Trust)*

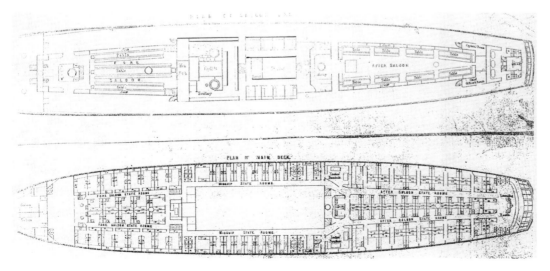

LEFT A plan of the cabins after the 1852 refit. *(State Library of Victoria/ss GB Trust)*

classes which will be more comfortable than 500 or 600.'

W.D. Waters' arrival on board was bewildering: 'I felt quite at a loss to know where to go. At last I was shown my Birth [sic] by one of the stewards.' The departure from Liverpool might have been highly emotional, as many of the emigrants knew it was unlikely they would ever see their homeland or families for many years, perhaps never again. But the accounts in diaries and letters are usually very restrained, Victorians did not like to show emotion. One passenger wrote in 1871, 'I delight not in trespassing on the sacred field of sentiment, as we have it displaid in the interchange of farewells. . . .' Some, such as Rachel Henning, were looking forward to joining family already in Australia.

Some were by now seasoned travellers; and many of the emigrants perhaps had little conception of how far they were to travel.

Robert Saddington at least had the pleasure of waving goodbye in 1853: 'Looking out of the port hole when one of these Steamers was passing I saw to my great delight my Father, Sister, Aunt and Miss H. To attract their attention I waved my Screw Driver – shortly after this I heard the Cry of Dinner.' Despite bands playing cheerful music and the cheers of people on board and ashore, Mrs Park struggled to conceal her feelings. 'She could hardly stand to look at the shore and the scene that was taking place so near the ship.'

Cabins

Cabins were very small by modern standards, and the bunks were even smaller. On her honeymoon voyage in 1866, Mary Crompton described and drew her 9ft by 6ft cabin.

No 1 is the berth. Joe has had the lower one enlarged and we only use the top one to keep things in, such as our warm rugs, coats dresses, bandbox etc. No 2 is a sofa all nicely padded, over it is a shelf with a ledge where we keep all our books, and a candle lamp is fixed to the side of the cabin. No 3 is the door which opens into a little passage; no 4 is the washstand, between it and the bed we have placed one of my boxes which serves as a

seat, all the rest of our baggage is put under the berth and sofa and firmly cleated down.

Rachel Henning found it difficult to share a cabin with a stranger in 1861:

I liked some of my fellow passengers very much, but not my cabin companion, Mrs Bronchordt. At least I did not dislike her, for she was good natured enough, but a more helpless, childish, complaining mortal it was never my misfortune to come across. She could not do her hair fit to be seen, nor mend her clothes, not keep her things in any sort of order.

The walls were thin and every sound was heard. When this was combined with a difficult passenger, life could be truly miserable, as reported by the Henning sisters:

Biddulph was very disconsolate with his fellow passenger, an old man who goes by the name of 'Pickwick' on board, and is so oppressed either by sea-sickness, or his conscience or something, that he does nothing but groan all night to the total prevention of sleep, either of Biddulph or the inmates of the adjoining cabins. They testify their disapprobation by knocking on the walls continually, the only effect of which is the increase of noise and ejaculations of, 'Oh dear what will become of me? Oh dear what shall I do. . . .'

Food

ABOVE **Inside the reconstructed galley on the *Great Britain*. On a typical voyage the galley was staffed by ten cooks, two bakers, two butchers and a storekeeper.** (Paul Blakemore)

RIGHT **A single cow provided fresh milk for the provision of passengers during the voyage.** (Author)

RIGHT **Some 500 chickens provided eggs and meat.** (Jonathan Falconer)

Feeding several hundred people for two months without the aid of refrigeration was a huge task. On the notorious ninth voyage, the chief cook was 'very bad, not equal to his work, looking dirty'. The second cook did all the work, and his dishes were 'very much praised in the After Saloon'. By 1870 Elizabeth Joseph found, 'Each class has a cook. He is very fine. There is one or other making tarts or pies every day. It gives a deal of trouble. He is not bound to do it. They are going to make him a collection.'

In 1862 Mr Dearlove recorded the animals brought on board before the voyage:

1 Cow is kept for the use of the First Saloon. On leaving port the carcasses of several Bullocks, pigs, sheep, calves slaughtered are taken on board and cart loads of all kinds of vegetables. Also alive about 3 bullocks, sheep 150, pigs 30, fowls 500, ducks 400, Geese 100 Turkeys 50, and for their support about 4 tons of Hay, 2 tons of Mangold

Wurtzel, 1 ton turnips, 1 ton oil cake, 1 Ton Bran, 5 sacks of peas and about 120 Bushels of grain, and last not least the refuse from the tables of 700 to 800 people. A perfect paradise for Pigs.

ABOVE LEFT AND RIGHT **There were storerooms for dry and tinned goods, vegetables, cooked meat and cheese.** *(Jonathan Falconer)*

FAR LEFT **Fresh fruit was issued when it was available.** *(Paul Blakemore)*

LEFT **Fresh meat and fish for the galley was butchered on board.** *(Jonathan Falconer)*

One young lady came to love them and was distressed when they were slaughtered for food. 'My daily treat was a visit to the various pens. Deeply I wept when I discovered the absence of any of my pets.'

The quality of the food varied greatly according to the class and tastes of the passenger. To Anne and Amy Henning in 1853, 'Our breakfasts are wonderful in the way of meat. There is generally hot beef steak, mutton chops, boiled salt fish, herrings, potted anchovies, cold ham, tongue, beef and boiled eggs, eggs and bacon, cold boiled fowl, hot rolls, tea, coffee, toast.' But to their sister Rachel in 1861, 'There is a quantity of food, but it is coarse; great joints of pork underdone and mutton and chiefly cold.'

Elizabeth Joseph described the second-class diet in some detail in 1879:

We have coffee for breakfast every morning. We have dinner at half-past 12, pork and preserved potatoes. The pork is very good but I don't care for the potatoes. . . . I take bread with it and make a little jam tart now and then. Other times we get beef, rice

LEFT The dining saloon was also used in between mealtimes. Passengers made their own entertainment, for example with the round brass nut-spinner seen here. It may have been used for gambling, but its precise use remains a mystery. *(ss GB Trust)*

ABOVE First-class passengers came to the dining saloon to enjoy the best that the ship's galley could offer. *(Alistair Hood)*

LEFT *Great Britain* saloon plate. *(ss GB Trust)*

"GREAT BRITAIN" STEAM CLIPPER.

67th Annie August 2nd 1861

BILL OF FARE.

Dishes.		Roast.	Boiled.	Dishes.	VEGETABLES, ASSORTED.
	Pea Soup				
	Salmon Fish	1	6		Pastry. Tartlets
3	Pork Apple sauce		6		Plum Pudding.
	Beef Corned		6		Rice do.
1	Mutton		3		Suet do.
2	Veal Curried Mutton				Bread and Butter Pudding.
	Turkeys				Roll do.
	Geese Apple Sauce				Custard Cabinet
1	Ducks	1			College do.
3	Fowls	2	1	6	Apple do.
	Chickens				Apple Tarts.
6	Mutton Cutlets Roast				Fruit do.
	Veal do. Roast				Open do.
	Stewed Steaks				Omelettes.
	Fricasses of Fowl				Maccaroni.
6	Currie au Vent Lobster				French Pastry.
	Tripe				Stewed Prunes.
1	Ham				
	Tongue				
1	Pork				
	Mutton Pies				
1	Pigs' Head				
	Haricot				

DESSERT.

Figs Raisins Almond Walnuts Bar Nuts

ABOVE This bill of fare from 2 August 1861 lists some of the dishes served to first-class saloon passengers, which includes such widely varying fare as corned beef, duck and apple sauce, and lobster vol au vent.
(ss GB Trust)

RIGHT Between the decks on an emigrant ship, 'Feeding Time' by A.B. Houghton, 1870.
(TopFoto)

and plum puddings. We get fresh bread three times a week in little loaves – Sunday, Tuesday and Friday. I can get anything here in the store. The bread gets short every week. I am only sorry I did not bring flour with me. It is very dear here . . . I have bought Tea, Flour and Sugar. I don't like the ship's tea at all.

Steerage passengers such as Walter Parr had some role in organising their own food.

I received a promotion this morning having been made Captain of my mess no 5. Consequent on my promotion I was disagreeably surprised to find that I was obliged to get up at ½ past six in the morning for the purpose of receiving provisions for the week. I had to undergo another dreadful crushing. But eventually I succeeded in getting the provisions which on the whole are very good as far as I am concerned.

On deck

The different classes lived separate lives, below decks at least. As *The Great Britain Times* reported: 'The first cabin knows nothing of the second – the second of the third; and

in many instances the passengers of their respective classes know but little of each other.'

On deck it was often difficult to keep the classes separate. In December 1867, 'We the second-class passengers sent a letter to the captain today asking him to keep the Emigrants from that part of the ship allotted to us these people are continually coming upon our Deck, and most of them have such an unclean appearance that we are afraid of having upon us what we should not otherwise have.' But as a steerage passenger in 1870, John D. Parr saw the matter very differently when there was a complaint by 'Three Stuck up Females . . . and a hook nosed half swell individual . . .'. As Parr saw it:

> Up to the Present time we have been permitted without giving offence to promenade any part of the decks but that directly Aft which is occupied by the Saloon

and to disalow this privilege now would discommode the Steerage and Intermediate passengers considerably and we were determined on appoint of the many Female passengers that are among us not to submit to this paltry line of distinction without a desperate struggle.

Parr retaliated with cartoons of the 'Hook nosed petitioner', which got him into further trouble.

First-class passengers were allowed to go anywhere on deck, which had the potential for some social mixing. Particular groups would probably find their own areas, as described in the ship's newspaper:

On deck a party were enjoying themselves keeping time to the merry notes of a flute, which found vent in the lively strains of an Irish jig. A party of Welch people, of whom there are a great number on board, were chanting forth their congregational hymns. A stray group were engaged shuffling the pack, anxious of possible to turn up a trump; whilst a few were seeking solitude and isolation, thinking, doubtless, of 'absent friends and memories dear'.

But the deck of a sailing ship was a dangerous place for the uninitiated. 'On Wednesday a young man in the steerage, while holding one of the sails, was lifted suddenly up, and, falling on the lower deck, had one of his feet severely sprained.' Even worse happened to Charles Wortman, a 6-year-old steerage passenger in 1865:

. . . the child was led from curiosity to approach the aperture, through which the windsail passes to convey the cold air to the Engineers, &c., below, when suddenly slipping fell through the hole on to the engine, about 21 feet below his head coming into contact with one of the cylinders. One of the engineers being close at hand immediately dragged the child from its perilous position, when surgical assistance was splendidly rendered, but all to no purpose. . . .

Religion

Though the ship did not carry a chaplain, clergymen of various denominations were usually to be found on board and might be offered some privileges for conducting services. In 1861 a Mr Curry was promised an upgrade to the saloon, but that was not forthcoming and he refused to conduct services until a subscription was raised among the passengers to pay for it. Nearly all the passengers and crew were Christians, apart from a group of Jews in 1853 who, according to the Henning sisters, 'had a grand assembly, both tonight and yesterday evening, in one of the Boudoirs, to celebrate the commencement of the year'.

Religion tended to divide the Christians as much as it united them. 'We had the usual services on today, viz. the Anglican Church Service in the Saloon, the Scotch Church in the

ABOVE The line on the deck that was not supposed to be crossed, except by first-class passengers, with one of the ship's guns. *(Jonathan Falconer)*

Second Cabin and the Catholic Service in the Steerage,' reported D.W. Waters. Observance of the Sabbath was important in marking the passage of time, though the landsman's concept of the 'day of rest' was irrelevant at sea.

Mary Compton was not happy with the 'dreadful old stick of a clergyman' on her voyage of 1866. 'He reads very very badly with a nasal twang, a sing song manner and [has] a peculiar habit of raising his voice at the end of each sentence.' A few weeks later she noted the incompetence of the organisation. 'Sunday, June 17th Mr Conabee preached, he had not arranged the hymns with anyone, so Mr Murray and Dr Sanger each waited for the others to begin, when up jumped three men from the second saloon and began to sing a most dismal tune very slowly.' Robert Saddington had a much more satisfactory experience in 1853:

This is the first Sabbath I have ever spent at sea and this day has it been manifested to me that my God confines himself to no place or home of worship. . . . In the morning notices were posted announcing that Service would be held in the After Saloon at 10.30 & 8. Another notice appeared stating that the Wesleyan Congregation would meet in the Fore Saloon at 10 & 7. Passengers were invited to attend accordingly I and Mr. Smith Repaired to the Fore Saloon at the appointed Hour when Mr. Davies a Wesleyan Missionary to Melbourne commenced by singing, this was followed by a beautiful prayer – he then

Law and order on the high seas

The *Great Britain* did not normally carry police, so the ship's officers were responsible for keeping order, with the assistance of the crew and occasionally the aggrieved passengers. It was not always an easy task.

The first Australian voyage attracted numerous gold prospectors who were not the most law-abiding citizens. At one stage two passengers were put in irons for drunkenness and disorderly conduct, and a second cabin passenger was accused of theft. The situation was described by Olcher Fedden:

12 passengers as jurymen were chosen 4 from the Aftersaloon . . . 4 from the Foresaloon and 4 from the 2nd cabin. Friend Duffet being a solicitor was chosen to prosecute and a Mr. Hodgetts of the same profession was chosen for

the defendant. The investigation lasted more than 2 hours, but nothing could be proved against him but that he was drunk last night and had struck the meat with a penknife.

Things were generally quieter on later voyages, though there were occasional fights and thefts. In 1867 the captain had to threaten some of the passengers several times, and finally one of them was put in the 'black hole' after making a 'fearful mess' over a shared cabin:

This is a hole at the bottom of the ship 3 feet wide by 6 feet long, perfect darkness prevails and what is far worse the dust from the Coals gets into the place and makes it very dirty for when the man came out he was like a Coalheaver.

BELOW Marine officers' cabin. *(ss GB Trust)*

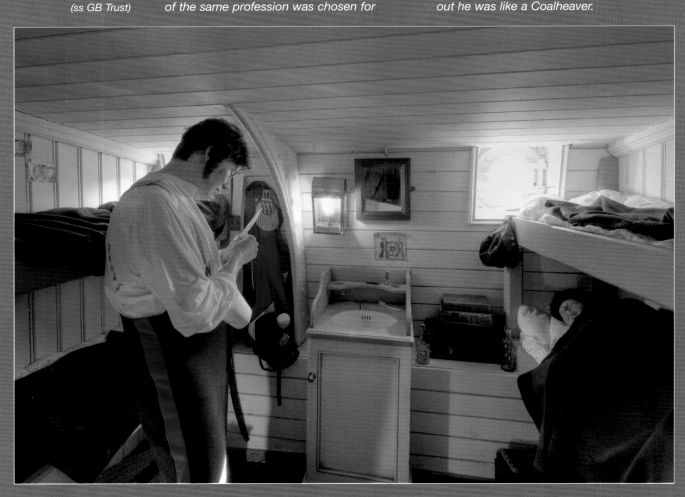

read a Chapter in Matthew and after Singing delivered a short address. . . . This evening I felt it a privilege to attend Mr D.'s Ministry. . . . His address was impressive and eloquent and his prayers were indeed fervent and I much enjoyed the service. Our Saloon was full and I noticed some from Aft.

Romance aboard

Love affairs between passengers were mostly discreet, but in 1872 Luisa Buchan noticed, 'some of the young ladies and gentlemen were very intimate indeed'.

Anne Henning, reported, 'One of the midshipmen, Mr Swan, always dances with Miss Middleton. Her mother found it out and did not like it and has forbidden her to dance at all. She cried and Mr S. declined dancing with anyone else, but Mrs Middleton was inexorable and there the matter stood.'

Later two passengers became engaged, but annoyed the others by monopolising the music room, until they took action: 'We had some music afterwards, not for the love of it, but for the sake of driving Miss Fitzgerald and Mr Hutchinson out of the room . . .'

In 1862, a correspondent of the ship's newspaper, *The Cabinet*, suggested '. . . it would be much to the young ladies' advantage if they could keep themselves free from the fetters of the winged god till they are once more on land, when they will have a wider field to pick from'.

Another replied, 'Because a gentleman is attentive to a young lady, forsooth they are engaged! . . . If any of my fair friends have been so foolish as to bind themselves, I would advise them to unloose the cord gently, and get it broken entirely before we reach Melbourne. . . .'

Medicine and hygiene

The surgeon was an important figure on an Australian emigrant ship, with equal authority to the captain over the condition of the passengers. Ship's surgeons were not well regarded by the rest of the profession and in 1881 it was reported that, apart from a small number of 'men of mature age and experience, who have adopted the sea as a profession,

and have stuck to it for years', they fell into two main categories:

There is first of all the youth just qualified; he may be possessed of a good deal of book-learning, and a good deal of dissecting-room lore, but he is almost necessarily ignorant of the routine and the requirements of life on board ship. . . . The second class consists wholly . . . of the 'bad hats' of the profession . . . dull, drunken and stupid.

Certainly John D. Parr had no respect for Surgeon Puddicombe in 1870. He was known as 'the Cowboy Captain' because of 'his "Youkel" like face figure and appearance which would in any other dress than the uniform of Dr pronounce him a farm Servant'. He was also suspected of entertaining women in his quarters late at night.

Others were of much higher calibre. Andrew Alexander served in most of the early voyages and again in the 1860s. According to the passengers in 1864, 'No matter at what hour, or the number of times he may have been summoned, he has always evinced the utmost good nature and benevolence to all classes, without respect either to their social or pecuniary position.'

Samuel Archer had trained at Liverpool Royal Infirmary and joined the ship for the trooping voyage to India in 1857, staying on for the next voyage to Australia. He eventually became a Surgeon-General in the army. Charles Morice was born in Wales in 1834 and studied at Guy's Hospital, London. He made five voyages between 1858 and 1861, and in September 1860 he was given purses of gold sovereigns by passengers in recognition of his services. Thomas Hocken trained at Durham and Dublin Universities. He became a ship's surgeon for the sake of his health and made two voyages in 1861–62, when he contributed regularly to the ship's newspaper.

The surgeon had to deal with bad weather and sudden changes of climate, often with passengers who were mentally unprepared or had not brought suitable clothing. On the way to New York, the ship often had to steam against the prevailing winds in great discomfort. On the Australian voyages the passengers and crew had to cope with a great variety of weather, from the equator almost to the Arctic Circle.

Rosamond D'Ouseley complained, 'The heat is now beginning to be perfectly frightful, in our cabin we can hardly breathe it is so stifling, all the light dresses are coming out (my buff print among them!) great swells are appearing & new fashions particularly in the saloon.' To Olcher Fedden in 1852, 'The perspiration stands upon one's forehead in drops, though taking no exercise.' Conversely, John Gurner reminisced:

I have in my memory the cold experienced during the three weeks we were in the depths of winter, south of the fiftieth degree of latitude. For some days we were nearly fifty-eight degrees south of the equator, and all that time running before a fierce westerly gale, carrying all plain sail, mountainous seas behind us, passing several ships hove to.

Bathing was not universal in the 19th century, but baths were available for the first-class ladies in 1853. The Henning sisters found outrage when '. . . a bath, having been pumped up by the cabin boy for a lady, while she was coming, one of the Fore Saloon passengers quietly stepped in, locked the door and took it for himself'. Not only was he from the wrong part of the ship, but men were expected to pump the bath themselves for ten minutes, and '. . . as some think, not worth the trouble for the sake of the advantage'. In 1869 Rosamond D'Ouseley got up early for a shower bath on deck, but 'there were too many gentlemen about – in future we intend trying it before 5 a.m.' In 1862 Mr Dearlove was one of the gentlemen in question. He got up at 4.30 but had 'great difficulty to get under the hose for a minute after waiting so long and depriving oneself of their sleep . . .'.

The Victorians were particularly prone to seasickness. They were the first people to

BELOW A typical Victorian view of seasickness, by the cartoonist John Leech.

steam directly into the wind, which could make a ship very uncomfortable, but they had no pills or stabilisers to remedy it. It was usually worst at the beginning of a voyage. Mary Compton overcame it but noticed that 'I was the only lady at breakfast, there was only one besides myself at lunch. I felt rather green but was not at all seasick. Poor Joe was miserably sick.'

Surgeon Hocken wrote that the disease was 'not considered of sufficient importance . . . and consequently, like toothache . . . it exacts no commiseration, but is rather considered legitimate food for mirth'. He recommended fresh air – 'Let one seek refuge in his berth, and most assuredly he will be a member of the "dead men" for days to come.'

Births were quite common on board and usually passed without much comment. Deaths among passengers and crew were equally common with three or four dying on the average voyage. Surgeon Hocken was philosophical: '. . . Death has been busy amongst us – two infants have died from measles and croup, and a poor suffering man from consumption. . . .' This was the fourth on the voyage, but not uncommon 'among such a large number as 700 people'. Deaths were public affairs and funerals were held, as described by John Parr:

> His Burial Service took place tween decks at ¼ to 10 Oclock this morning. From the main deck I saw his canvass covered Body launched into the foaming waves which seemed to leap upwards to receive the Blood Staind and heavily weighted carcase. It sank immediately leaving a faint Blood stain upon the wave that received it which in a second was hurled from our view. . . . The Funeral service was led by minister passenger and as many Sailors as could be spared from their duty were present.

But sometimes the sadness did not last long. After the death of a child in 1867, W.D. Waters reported, '. . . it was sewn up in canvas and the burial service read over the Body and then cast into the deep. There was dancing again this evening on the poop Deck.'

Letting off steam

Saturnalia and ceremony were important to seamen. They were paid a month in advance and liked to believe that they were working for nothing during that time, so they celebrated the end with the 'dead horse' ceremony, as described by John Campbell in 1871.

> . . . they make a temporary horse (not very artistically got up) and they dress fantastically, and go marching round the deck singing. One on the said horse, and his bodyguard with him, after some singing and marching to the amusement of all (and the dread of some), the man and horse are pulled up, with a tackle, to the yard arm. After three cheers, the poor old horse is dropped into the sea. As I see him floating past, I think to myself a very true representation (in many cases) of how the sailor has thrown away his advance money, that he has been working for the last month.

The dead horse ceremony only involved the sailors, but crossing the line, or the equator, affected everyone on the Australia run. It was anarchic, riotous and sometimes dangerous, especially in 1852 as described by Olcher Fedden:

> Mr. Gray, the Second Mate, was dressed up as Neptune with his trident. He wore fearful mustachoes and long grey beard and in costume admirably represented his Majesty. His wife also accompanied him, this was the Boatswain dressed up. The ceremony is by no means a form. All the sailors and young midshipmen were shaved by him. His emissaries, about 10 in number with their faces blacked or otherwise disguised, go forth and bring them one by one to him. Resistance is out of the question, it is surprising with what determination they go to work, nothing will induce them to release one. . . . Most of us escaped by making a subscription for the men, but those who would not give, or those who gave stingily did not escape. Those also to whom a grudge was owing, namely the old man who had said he was robbed of his money, got it most unmercifully.

The Great Britain Times.

VOL. I.—Nos. 1, 2, 3, 4, & 5.]   "MAGNA EST VERITAS ET PRÆVALEBIT."   [Price

"GREAT BRITAIN" STEAM SHIP, TABLE BAY, THURSDAY, OCTOBER 14, 1852.

COPY OF THE "GREAT BRITAIN'S" LOG, WITH ADDITIONAL NOTES.

THE LATITUDE AND LONGITUDE, WITH DISTANCE RUN, ARE KINDLY FURNISHED BY CAPTAIN MATHEWS.

Journal of Daily Transactions on board the "Great Britain"

ABOVE The first page of the *Great Britain Times* from 14 October 1852. *(ss GB Trust)*

Ship's newspapers

On many voyages, the passengers produced a ship's newspaper. On starting *The Albatross* in 1862, Colonel Sir James Alexander explained, 'On previous long voyages, having found the advantages of a weekly paper to afford a little agreeable excitement, to stimulate latent talent among the passengers, and diversify the monotony of the voyage.'

There were no printing facilities except in the voyage of May–August 1861, when a passenger had a small press, but the results were considered disappointing. Otherwise it was necessary to read the paper out at an assembly, or to make copies by hand. In 1862 the editor of *The Cabinet* advertised for 'any

number of nice young men who can write a plain round hand, and are ambitious of enjoying the improving influences of intellectual society'. Three handwritten copies of the *Great Britain Times* were to be produced in 1865, one for the saloon passengers, one for second class and a third for 'third cabin, steerage and sailors. The latter to have it one day.' But in December the editors complained that 'It is very evident, from the fact of the Second Cabin and Intermediate copies of our last publication having been either thrown overboard, or otherwise destroyed, there are several parties on board who are extremely anxious to crush our humble periodical. . . .'

Some newspapers were published at selected moments in the voyage, such as rounding the Cape and crossing the line. *The Albatross* was published weekly while *The Vain Effort* was far more ambitious in trying to keep up a daily edition.

Each edition had an account of the ship's progress provided by the officers, including latitude and longitude, distance run, work of the engines and weather conditions. Many of the newspapers used very weak puns of a type that appealed to the Victorian sense of humour, often based on nautical terms: 'When does Captain Gray, when giving an order, lead you to suppose he is going to take part of the ship away? – When he orders her "half astern".' Most included reminiscences of passengers, and indeed these dominated *The Albatross* of 1862. *The Cabinet* of the same year was probably the most informative and published articles on the distinguished passengers including 'the eleven of all England', a satire on the Irish which is offensive to modern sensibilities, poetry, natural history, letters from the public, and spoof advertisements.

Some ships' newspapers were printed after arrival in port, but during the fraught voyage of 1852 the captain went to some length to prevent a printing during the stop in Cape Town. *The Vain Effort* of 1860 was 'published by subscription (so general as to merit the title "universal") of the saloon passengers . . .'. An alternative to a regular newspaper was to print an account of the voyage as a memento for passengers, as Gus Hatton did with *The Great Britain Chronicle* in 1864.

Entertainments

Passengers had to get used to a rather monotonous life, like Mary Compton.

It is a queer life here, it is pleasant enough but very unsatisfactory. I seem to get nothing done what with eating and sleeping there is very little time for anything else. We are never up before 9 o'clock (breakfast time), it is very lazy but there is no use getting up, the deck is so wet one cannot walk, the saloon is very cold and the children are having their breakfast and it is not pleasant doing anything in our cabin till it has been aired. . . .

Many of the passengers organised their own entertainment in true Victorian style. Captain Gray encouraged such ventures, for he was 'always glad when the people enjoy themselves and take plenty of recreation'. Anne and Amy Henning helped to organise a concert in 1853.

Dr Sicler opened the concert. We had some nice playing and singing. Mr Stevens, the American, was the lion of the evening and was encored in both his songs. It is certainly a great treat to hear him, even to one who knows as little about music as I do. There were refreshments handed round between parts; everything was made by the confectioner on board and was as good as you would get at a party in England. We had champagne and champagne punch and plenty of other good things. The concert was over about ten and then the benches were taken away and we had a dance to conclude the evening.

Mary Compton took a leading part in one concert herself, but was scathing about other people's efforts: 'The second class gave a concert on deck, it was much too long and a very dreary collection of songs.' And three days later, 'We had another concert on deck, really I should think there was never a ship so full of persevering musical talent.'

Amateur dramatics were surprisingly rare, but in 1860 the passengers put on a production of *The Merchant of Venice*, which according to *The Vain Effort*, 'merits something more than a passing notice, not only from the amount of time spent on its "getting up", but because so many are engaged in it'. It was marred on the first night by 'loss, of memory' by one actor which put the others off their stride. In 1862 some of the passengers staged the trial scene from *The Pickwick Papers,* which was watched by at least 500 people, 'many betaking themselves to the rigging after every available stand-point was occupied, until the court-room formed the centre of a perfect amphitheatre of faces'.

Some of the more learned passengers gave lectures to 'alleviate the tedium'. 'The subject of the lecture was the proverbially dry one of Law; but, treated as it was by Mr. Wayte, who with good taste interspersed his text with many amusing anecdotes and *bon mots*, the audience were both instructed and entertained.'

Some took an interest in the wildlife. In 1868 the chief engineer's son caught an 'Albertrous' by trailing a fishing line astern of the ship, though in 1862 it was said that Captain Gray 'wont allow any albatross to be caught because he says it will bring on a gale of wind'. Passengers often got excited by marine life, for example when '. . . about 11 o'clock everyone rushing to the ship's side to see a shoal of porpoises coming close to the ship, & immediately after we lost sight of them a whale was soon coming towards us spouting water up to the height of about 3ft . . .'.

In 1870 John D. Parr saw '. . . a large concourse of passengers curiously gazing through Burnt glass pieces of Broken Bottle, the

LEFT 'Cock fighting', from *P&O Jottings*.

ships Sexton and a variety of other mediums, at an eclipse of the Sun'. He claimed that his own eyes were 'wonderfully strong' and he did not need the 'above mentioned agents' but one wonders how he avoided damage after 'gazing on it for some time'.

Passengers were often to be found reading on deck and in 1853 there was a small ship's library, while women sewed. In 1861–62 the England cricket team practised on deck, and also indulged in a game called 'cock fighting' which was 'played between two opponents, who had a stick passed at the back of their knee joints, and held it there with their arms. While in this position each had to knock the other down.' On another occasion one of the passengers used a belaying pin as a bat, but it slipped from his hand and injured another. Often there was gambling, and in 1865 Captain Gray '. . . sharply reproved' the steerage passengers 'for not taking more open air exercise and keeping their berths clean'. Gambling tended to die down towards the end of a voyage. 'There is not so much card playing today. Those who have won do not wish to lose and the losers have little more to part with in some cases. One man has won upwards of £200 since he has been on board,' reported the Henning sisters.

Volunteers

Though merchant ships no longer carried whole decks of guns as they had done in the age of war and piracy, the *Great Britain*, like many merchant ships, originally had her sides painted in the black and white 'Nelson chequer' with imitation gunports on a white band. That was purely stylistic, no one could confuse her with any warship, but she did carry some real armament, and two 12-pounder carronades

were among the goods sold after her standing in Dundrum Bay. These were relatively light, short-range weapons mainly for signalling and salutes, but on the Australia run she often carried up to ten cannon, to protect the gold she was carrying back.

She was of course a merchant ship, but a large proportion of her officers and crew were in the Royal Naval Reserve after it was founded in 1859. According to Felix Grenyer (an RNR man himself), two-thirds of the crew were members in 1867–68. In Liverpool they trained in the drill ship *Eagle* and were supposed to have done their four weeks before signing on for a long voyage,

'. . . but Capt. Gray was the only Officer in the Royal Naval Reserve who was allowed a relaxation of it in his favour. If a Naval Reserve man had done 2 or 3 weeks drill for the current year only and wanted to join the "Great Britain," he was allowed to do so if Capt. Gray wanted him.'

Apart from trooping for the Crimean War and the Indian Mutiny, the ship was caught up in scares. In 1862, when there was a possibility of involvement in the American Civil War, the passengers organised a volunteer corps for defence against privateers or warships, led by a former sergeant nicknamed 'General Gumbum' and described by Mr Dearlove:

. . . we had the Volunteers headed by their drill sergeant with his coat turned to show the red lining; also a flute and tin Bucket for a drum, which constituted the band, and the ensign carried a flag made of one of the old numbers of the ship's signals, and they were drilled and marched much to the amusement of everybody.

Another corps was formed in 1864, with more dramatic results, according to Thomas M. Dunn:

Tuesday 24th was memorable on account of the battle of the 'Great Britain' fought that evening. Colonel Jacobs of the volunteers and Captain Bruce of the same corps marched their men with rifles right round the deck. When passing the forecastle the sailors twice opened fire upon them. . . . Fire was again opened upon them. They immediately stormed the outworks & attempted to take the battery & was gallantly defended by the sailors under two leaders with basket sticks & an old pocket handkerchief on a pole for their flag. A hand to hand melee took place the advantage was decidedly on the side of the sailors. . . . The battle was waging hot & strong & threatened to be very disastrous to the combatant, when the chief mate appeared on the scene and put an end to the fight. . . .

The end of the voyage

The passengers began to sense the end of the voyage was near when the crew started to make greater efforts to clean and paint the ship.

The sailors are very busy washing the vessel because it has been painted in and out with drab paint – that is – all the boats railings that is round the ship, the engine house and the forms we have to sit on. The stewards are downstairs scrubbing the cabin doors and all the passages and windows.

There was invariably a series of grand dinners during the last days. Mother Mary Paul Mulquin declined to attend for religious reasons in 1874, but looked in on the preparations. 'We visited the saloon at 9.30 – the hour appointed for supper, and got a full view of a scene of fashion, and most exquisitely arranged tables . . . decorations hung around and viands of every variety covered the tables – champagne in abundance.' But others took it very seriously. It

. . . created great commotion particularly among the ladies who ransacked their wardrobes to appear in their best for the occasion at length the appointed hour arrived when a good substantial dinner was served up after the tables were loaded with a great variety of fruits and wines, a chairman being appointed a great number of Toasts were proposed and drunk with great enthusiasm among those who stood first as it deserves to be on the 'Great Britain' the health of the Queen, and Royal family, next came our worthy and excellent commander John Grey which was drunk a 'hip hip hurrah' . . .

During the notorious first voyage to Australia, Felchen had difficulty raising support for a testimonial to the captain. 'I am sorry to say I found a very bad feeling towards him, and very many refused stating that he had been far from kind and courteous to them.' Eventually he persuaded all but 10 of the 80 after saloon passengers to sign. More typical was what the second-class passengers wrote of Captain Gray in 1865.

As a seaman, we have found exhibited by you the utmost zeal and skill in taking advantage of every moment and circumstance to expedite our arrival at our destination. . . . As a gentleman, your urbanity to all classes, your unrestrained intercourse with passengers in every part of the ship, and the regularity and undeviating good-humour . . . has established you among our recollections of the voyage as our best friend on board.

Passing ships had usually been spoken to during the voyage, but they rarely had up-to-date news. In August 1861 the Liverpool pilot brought a newspaper aboard and it was read to the passengers, causing the Americans to be distressed about the Civil War in their country. In April 1862 the passengers and crew heard from the Liverpool pilot that Prince Albert, who had launched the ship, was dead. And in August 1870, passengers heard that 'some dreadful battles had been fought between France + Prussia within these few days and that Both nations were now engaged in a War as Yet unequalled for Blood Shed'. Many passengers expressed regret at leaving the ship. Rachel Henning wrote, 'I am very sorry to say good-bye to the Great Britain and her captain. . . . She is a splendid ship, and I am sure we all have reason to speak well of her.'

BELOW Passengers' luggage stacked ready for unloading. *(Jonathan Falconer)*

Chapter Seven

Conservation and restoration

After lying abandoned for 33 years in the South Atlantic, the SS *Great Britain* was refloated and towed 7,000 nautical miles home to Bristol from the Falkland Islands on a pontoon in a pioneering salvage operation. But returning the ship to her birthplace in the Great Western Dock was only the beginning of an epic conservation story.

OPPOSITE The hulk of the *Great Britain* at Sparrow Cove in the Falkland Islands in 1969. The three 1857 original lower masts and the iron mainyard survive from her sailing days, and are now in the museum in Bristol. *(South American Pictures/Marion Morrison)*

In the Falklands

After her insurance write-off and sale in 1886, the hulk of the *Great Britain* spent 50 years anchored off Port Stanley, serving as a wool storehouse for the Falklands Islands Company. An extra cargo door was cut in her starboard side abreast of the main hatch and the former funnel position, which weakened the structure almost fatally. She was still venerated by the islanders and visited by the officers of Admiral Sturdee's fleet when he fought a battle with the Germans in 1914. By 1933 her weather deck was so leaky that she was useless for storage and it was possible she might sink and block the entrance to Port Stanley. In 1937 she was towed 3½ miles to Sparrow Cove, where she was scuttled by cutting seven holes in her hull.

Recovery

The salvage and preservation of the ship was proposed as early as 1936, but £10,000–£15,000 seemed too large a sum in those days of depression. In 1957 San

ABOVE Ewan Corlett (left) surveys the ship with a member of the crew of the Royal Navy's ice patrol vessel HMS *Endurance* in 1968. *(ss GB Trust/HMS Endurance)*

BELOW Below the weather deck of the *Great Britain* were little more than rusting iron frames and girders. No wood decking remained and water in the bottom of the hull rose and fell with the tides. *(South American Pictures/ Marion Morrison)*

RIGHT Inspecting the main mast between decks. It had been hoped to raise the masts in their entirety but the heel tenons were solidly rusted in, so the fore and main lower masts were cut through with a chainsaw some 3ft above the main deck level. It was difficult moving around on the lower decks as there were only iron girders and a long drop to the water at the bottom of the hull. *(South American Pictures/Marion Morrison)*

Francisco Maritime Museum, under Karl Kortum, began to show an interest in taking over the ship but it was prepared to yield to a British effort.

The campaign might be said to have begun in earnest in November 1967 when the naval architect Ewan Corlett wrote a letter to *The Times*, still the standard method for launching a cause in Great Britain. He drew attention to the historic value of the ship and compared her to Nelson's *Victory* at Portsmouth and the clipper ship *Cutty Sark* at Greenwich, which were already preserved. 'Historically the *Great Britain* has an equal claim to fame and yet nothing is being done to document the hulk, let alone recover it and preserve it for record.' He suggested that the authorities should 'at best do something to recover the ship and place her on display as one of the very few really historic ships still in existence'.

No direct government finance was likely to become available, but with naval support Corlett was able to visit the islands in 1968 and survey the ship. After initial disappointment he concluded that the hull was salvageable despite many holes in her bottom, and a significant crack in her starboard side near the cargo port, which was widening and might cause the ship to break in two within a few years.

With finance largely provided by the patriotic millionaire Jack Hayward, a salvage expedition was prepared. For the second time in her long history, the *Great Britain* was to be refloated and towed away in a pioneering operation. Technology had moved a long way since

LEFT **Leslie 'Spike' O'Neill inspects the gap at the top of the starboard crack, which widened and dropped by almost a foot between the initial survey in 1968 and the salvage in 1970.** *(South American Pictures/Marion Morrison)*

ABOVE The first day pumping out the water. With the crack made watertight with mattresses and most of the scuttling holes sealed with plywood, diesel-powered pumps were started on 5 April to pump out water from inside the hull. *(South American Pictures/Marion Morrison)*

BELOW The German tug *Varius II* and the pontoon *Mulus III* are alongside the *Great Britain* ready for lifting the hull in April 1970. *(ss GB Trust)*

RIGHT Air is pumped into the pontoon and the bow of the *Great Britain* is lifted right out of the water for the first time since c.1882. *(South American Pictures/ Marion Morrison)*

Dundrum Bay in 1846, but the difficulties were still immense, as the ship had to travel thousands of miles across the ocean, through almost every climate. It was agreed that the hull on its own was not fit for an ocean voyage, even if the holes had been plugged.

The Anglo-German firm of Ulrich Harms and Risdon Beazley suggested that the holes in the hull should be blocked and the water pumped out to make her float. A pontoon would be sunk under her, then pumped out to take the weight of the hull. The pontoon *Mulus III* arrived on site on 15 March 1970, pulled by the tug *Varius*. There were many interruptions due to force ten storms in the high latitudes and great nervousness that the old ship would finally split in two during operations, but when the pontoon was floated under her on 13 April the ship settled on blocks carefully prepared from William Patterson's plans (published in John Bourne's *Treatise on the Screw Propeller*), and the crack and twist almost disappeared.

The ship was towed to Port Stanley for Richard Goold-Adams, representing the project, to take formal possession. She was then taken to Montevideo where she was further strengthened. She left on 6 May for the lengthy part of her voyage, more than 7,000 miles back to Bristol. Off Cape Finisterre she was sighted by an RAF reconnaissance aircraft and a photograph appeared in the press, marking the beginning of a media campaign.

Off Avonmouth, where she had waited for her trial voyages in 1845, she was taken from the pontoon and refloated for the last stage. The arrival of the ship in Bristol was delayed for a day due to bad weather, and on 5 July she was towed up the River Avon, watched by 100,000 people and filmed for national television. The view from Brunel's Clifton Suspension Bridge is always spectacular, but the sight of the old hull from above added an extra dimension. She lay in the Floating Harbour until 19 July, long recognised as the ship's birthday, when she was put back into the Great Western dock.

Getting the *Great Britain* back to Bristol was only the start of the story. Conserving a historic vessel is one of the most difficult and expensive tasks imaginable. There will inevitably

ABOVE Now afloat, the *Great Britain* is towed up the River Avon, 5 July 1970. *(ss GB Trust)*

BELOW She passes beneath Brunel's Clifton Suspension Bridge. *(South American Pictures/ Marion Morrison)*

ABOVE **Just a few of the 100,000 or so spectators who flocked to see her enter the floating harbour.** *(ss GB Trust)*

BELOW **Her return to the dry dock from which she was originally launched on 19 July 1843.** *(South American Pictures/Tony Morrison)*

be periods when the enthusiasm of amateurs and volunteers is not enough to cope with the demands. A ship is usually designed to last for around 30 years, and any time beyond that is a bonus. Unlike a building on shore, it is subject to the stresses of the sea, and there are many parts that are very difficult to inspect both inside and out. All this has to be done within a modern culture of safety, which is very different from the Victorian age. A captain like James Hosken, with his addiction to risk-taking, would not last long in today's merchant navy.

The story of the *Great Britain* is not without its mistakes and setbacks, but it provides many lessons on how ships can be restored in difficult conditions and eventually become important visitor attractions. But many factors need to come together for a project to be a success – the significance of the ship itself, and the public affection for it, finance, management, conservation skills, a sense of place, and interpretation. Apart from the historic value, none of these came automatically in the case of the *Great Britain*.

The *Great Britain* and Bristol

It was only by chance that the dock where the *Great Britain* had been built still survived, though with modifications and with changes to all the buildings around. The ship had originally fitted the dock 'like a glove', and it seemed only natural that she should be restored there and kept on permanent display. But to start with the City of Bristol was reluctant to accept this, for it had its own plans for the harbour and was aware of the expense of restoring historic ships.

According to the ship's newsletter of December 1970, 'We are told that the success of the salvage has itself made us rather like the self-satisfied cat, which comes and lays a dead mouse at its master's feet in the misplaced hope of being appreciated.' One councillor was overheard to say that 'the whole rust bucket should be sold up and the money used to build some more old people's houses'.

The Floating Harbour had been in decline for many years before it was closed to commercial

ABOVE The view over Bristol Harbour showing the restored masts of the *Great Britain*. *(Author)*

BELOW The Great Western Dry Dock with the *Great Britain*, viewed from the air in 2010. In the foreground is the floating harbour, with the River Avon at top left. The Great Western Dockyard buildings surround her original dry dock. *(ss GB Trust)*

Brunel's legacy

Despite dismissal by a rival spokesman as 'Bob the builder', Isambard Kingdom Brunel came second after Winston Churchill in a national BBC TV poll of the greatest Briton in 2002, six places ahead of Horatio Nelson. Brunel had powerful advocacy by the highly popular Jeremy Clarkson, but underlying that was the fact that the great engineer represented the status accorded to his profession in the days when Britain was great. The *Great Britain* is only one of several Brunel constructions in Bristol alone, alongside the Clifton Suspension Bridge and Temple Meads Station. Nationally, it has to share the glory with the whole of the Great Western Railway to Penzance, and the Saltash Bridge. But bridges and railways are too easy to take for granted and are often passed unnoticed. Nothing survives of the *Great Western* and very little of the *Great Eastern*, but the *Great Britain* can be seen today just as Brunel conceived her.

traffic in 1973, and there were even plans to fill in part of it in the late 1960s. Several alternative sites were considered for the ship, including Cardiff and Portsmouth, neither of which had any real connection with her. Liverpool was not mentioned, perhaps because it had found a new role in pop culture. St Katharine Docks in London might have involved a reversal of history, in that masonry would have had to be removed to get the ship in, but it was considered too risky. Reluctantly, Bristol City

BELOW Benefactor Sir Jack Hayward with a portrait of himself painted by Jerry and Anne Hicks in 1996.
(ss GB Trust)

Council agreed to accept the ship in February 1971, though it ruled out any possibility of financial support. In 1975 they gave the ship the lease of the dock at a peppercorn rent and its attitude began to improve.

Though the dock is some way from the city centre and transport links are not first class, the ship is now at the heart of Bristol. Her masts can be seen from some distance over the flat area round the harbour and tend to arouse the visitor's curiosity, as well as providing a focus for the rather disparate collection of heritage and leisure vessels in the Floating Harbour.

Finance

No preserved ship is financed by visitor receipts alone, and very few have direct funding from national or local government – certainly not the *Great Britain*. Sir Jack Hayward supported the ship in many ways from the recovery onwards, but even his funds were not infinite. The second major contributor was John Paul Getty Jr, the American billionaire who had long been resident in Britain. In 1983 he telephoned to offer £60,000, the full amount needed by a restoration appeal. He and Jack Hayward soon had a friendly rivalry with their contributions. In 1985 Getty proposed that they should give $½ million each, then they both raised it to pounds.

Another way of raising money is to let out space on the ship for corporate or private events. This was done on an informal basis until 1990, when 'SS Great Britain Banquets plc' was set up and the dining saloon was opened to host the events. But the high-quality restoration of the saloon proved very expensive, as much work had to be done on the ship's deck beneath it.

The *Great Britain*'s salvation would come from a different direction. The Heritage Lottery Fund (HLF) was set up in 1994 to help distribute the large sums of money being raised by the National Lottery, under strict rules. As well as charities, sport and the arts, 8% of its income was devoted to heritage. The organisers of the *Great Britain* project began to think seriously about a major bid in 1993 and submitted the bid in 1995. It was swiftly rejected, and the project was asked to reconsider completely what it was trying to achieve and how it could best do that. It would mean significant changes in its methods: to become more of a professional heritage operation, to focus on ship preservation and on engaging effectively with modern visitors, and to secure its rights to the extended site around where the ship was based.

A new bid had to be made in two stages and the Fund demanded complete detail of every part, which created a large amount of work. The Stage 1 proposal was delivered to London by train on Christmas Eve 1999. There was concern in July 2002 when the HLF wanted to defer its consideration, as it was uneasy about the innovative proposals on dealing with rust and wanted more time for peer-reviewed research. But by the end of the month the HLF had passed Stage 1, meaning that £7,300,000 out of a total cost of just over £10 million was allocated. Stage 2 was approved in October 2002. Although money was not allocated for restoring the engine and rigging, the future of the ship was assured.

Research

Much research was needed to restore the ship to her original condition, as very little apart from the hull survived. New scientific research was commissioned into the mechanisms of salt-driven corrosion in wrought iron in order to determine how the pernicious corrosion could be arrested without replacing the iron with replica steel. Even more research was needed to interpret the ship to visitors, to tell her story, to show how she worked and to represent the lives of passengers and crew on board.

The *Great Britain* had attracted a huge amount of interest at the time it was built, and during a large part of her active life. Brunel's own papers, in the British Library and Brunel Institute alongside the ship, are quite extensive but do not answer all the questions. Early in her life Christopher Claxton published a short but detailed *History and Description* of the ship, while articles appeared in the *Illustrated London News*, *Mechanics Magazine* and *The Imperial Cyclopedia of Machinery*. The annual reports to the shareholders of the Great Western Steamship Company give rather basic information on progress and changes in design.

In addition, there were many drawings, prints and paintings of the ship, though most of the original plans have not survived.

At the centre of the research for engine restoration was a set of prints published by John Weale of London in 1847. More general works by John Grantham, Robert Murray, John Bourne and Robert Spencer Robinson give information on the progress of steam engines, screw propellers and iron shipbuilding. They are mostly well illustrated and some include plans specific to the *Great Britain*.

In 1975 Ewan Corlett published *The Iron Ship*, deploying an impressive amount of research into the ship's history, and giving a personal account of its recovery – a book which has remained in print and is still difficult to better after nearly 40 years. Inevitably it was focused on the ship itself, rather than the stories of those who served or sailed in her.

Research in Australia, where the majority of *Great Britain* passengers travelled, has proved the key to the understanding of the ship in her heyday. More than 80 shipboard diaries have been found and many, such as those kept by Allan Gilmour and Mr Dearlove, give details of the sailing and routine of the ship. Apart from books and papers owned by the *Great Britain* Trust, most of the documents used in research were photocopied and are now available in the Brunel Institute attached to the museum, where they are accessible to researchers without having to travel the world to look for them.

ABOVE Dr Ewan Corlett and the *Great Britain* at Montevideo. Dr Corlett, the naval architect whose letter to *The Times* in November 1967 first prompted interest in the idea of salvaging the *Great Britain*, was in Montevideo to greet the flotilla when it arrived on 2 May 1970. *(South American Pictures/Marion Morrison)*

ABOVE HRH Prince Andrew visits the ship in 1983, with Ewan Corlett, Joe Blake and Basil Greenhill. *(ss GB Trust)*

BELOW The *Great Britain* maintenance team in 1997, left to right: Ray, Don Johnston, Martyn Bill, Frank Porter, Roger, Alan Johnson, and Gary. *(ss GB Trust)*

This is supported by the library of the maritime historian David MacGregor, which was left to the ship on his death in 2005. The papers relating to the ship are catalogued in Jean Young's bibliography, published in 2003.

As a result of the failed lottery bid the project had to become a more professional and museum-orientated organisation, and in 1997 Matthew Tanner was appointed as the first curator. A graduate in Maritime Studies at the University of St Andrews, he had worked in the Scottish Fisheries Museum at Anstruther and the Merseyside Maritime Museum. He played a key part in preparing the lottery bid and in developing the waterline sea concept, before succeeding Dr Bill Ellison as director in 2000.

Organisation

Soon after her arrival in Bristol, the management of the ship was taken over by the SS *Great Britain* Project, which consisted of hundreds of members guaranteeing to pay £5 each in the event of insolvency. This allowed the project to seek charitable status and become eligible for various grants. By 2002 the organisation had outgrown its remit and it was replaced with the SS *Great Britain* Trust. It continued to meet twice yearly and most of the work had been done by the executive committee, including Ewan Corlett and Basil Greenhill, formerly director of the National Maritime Museum in Greenwich. In 2008 the Trust's council began to meet more frequently, and the executive committee was abolished.

In the United Kingdom, it is usually necessary for a high-profile, non-profit-making organisation to find a high-profile patron, preferably royal. In the case of the *Great Britain* the role was taken by Prince Philip, a naval officer himself, the supporter of many maritime causes and, in a sense, the modern equivalent of Prince Albert, who had launched the ship in 1843. Prince Philip was succeeded by his son, Prince Andrew, also a naval officer, who became patron in 1983.

Early restoration pre-1999

It was decided in 1971 to restore the ship to her configuration as she left Bristol in 1845, with her original engine, six masts, and without the upper deck which was added later, but did not survive. This maintained the link with Brunel and with her launch in the city of Bristol, though it represented a period when the ship did not thrive commercially. The Australia run, on the other hand, took up the bulk of the ship's active life during which she was highly successful and popular. She carried 25,000 passengers there and back, and around a quarter of a million Australians can trace descent from them – a market which is catered for in Adrian Ball's book *Is Yours an SS* Great Britain *Family*? Moreover, the period yielded a huge resource of letters, diaries and memoirs compared with the almost barren New York period. In 1999 it was decided to represent this in the museum adjacent to the vessel, and in the reconstructions below decks.

The weather deck and the hull

In addition to routine and small-scale work, there were two main areas that needed a great deal of work and investment if the ship was going to be viable. The first of these was the weather deck, the uppermost one, which was exposed to the wind and rain. It was already in poor condition when the ship was salvaged from the Falkland Islands, and catwalks had to be used to give safe access. A secure deck was essential both to keep the hull dry and as part of the visitor experience. A new one was laid in the late 1970s, but within ten years it had seriously deteriorated.

In 'the most important addition . . . to the ship since she returned to Bristol', a steel deck was fitted in 1995. With a camber to allow water to run off and a bituminous coating, it protects the interior from water. To maintain realism, the steel deck is covered with Jarrah hardwood set 1⅛in (3cm) above the deck, thus allowing circulating air to preserve both.

The second major area was the hull itself, which presented the greatest challenge. The fabric of the hull of the *Great Britain* is exceptionally important because the frames and plates are mostly original. HMS *Victory*'s timbers and planks have nearly all been replaced during her long career, often in quite recent times, and the planking of the *Cutty Sark* was much damaged by fire in 2007. Moreover, the *Great Britain*'s structure represents a specific stage in the evolution of iron shipbuilding, and furnishes material for research.

Though iron proved remarkably strong, the ship saw hard service during her lifetime, both in 40 years at sea and, perhaps even more, in 84 years of neglect in the Falkland Islands. The repair of the hull is now strictly a conservation job, rather than restoration. Some areas had been repaired with steel plates in the 19th or 20th century, holes had been plugged up with fibreglass during the recovery in 1970, and after the wood sheathing was removed dome-headed rivets were used to block the holes that remained. Far more serious was the condition of the older plating. Only about 3% was in good condition, 29% was strong enough to withstand blast-cleaning treatment, 43%

LEFT Bob the blacksmith at work on deck. *(ss GB Trust)*

BELOW The deck prior to the start of deck laying in 1980–81. *(ss GB Trust)*

BOTTOM Deck laying nearly complete. *(ss GB Trust)*

had 'severe' corrosion and the remaining 17% had 'very severe' corrosion with parts missing, and 8% was covered with GRP, concrete or timber. Standard shipyard cleaning techniques had done nothing to stop the decline and new methods were needed.

Ten options were considered in 1999, including cleaning by normal shipyard techniques, coating with tannic acid, alkaline sulphite treatment, water diffusion of chlorides, electrolysis, and recording the ship and placing her in a sealed environment, and building a roof over her. The tenth option, environmental control, included three possibilities: to dehumidify inside a glass cocoon, to dehumidify under a glass roof, or to do so under a false waterline. This option was considered to be the most appropriate because damage was far worse in the lower part of the ship.

During her service the upper hull had been constantly maintained and painted, but while she was laid up in the Falkland Islands the lower part had been subject to the rise and fall of the tides and all the bad effects of salt water on iron. The new research demonstrated how dehumidification could stop the rot. An artificial water surface could be created in the dock, sealing off the most vulnerable part in a controlled environment at very low relative humidity to stop the salt-driven corrosion (under 20%). At the same time it would give the impression of the ship being afloat, while allowing visitors free access to the underwater hull to view its fine lines, unique plating, and the restored propeller and balanced rudder. This proved to be the stroke of genius that impressed the HLF, and defined the future of the ship.

BELOW A drawing from the conservation plan, volume 2, showing relative areas of iron, steel and fibreglass repairs in the hull. *(ss GB Trust)*

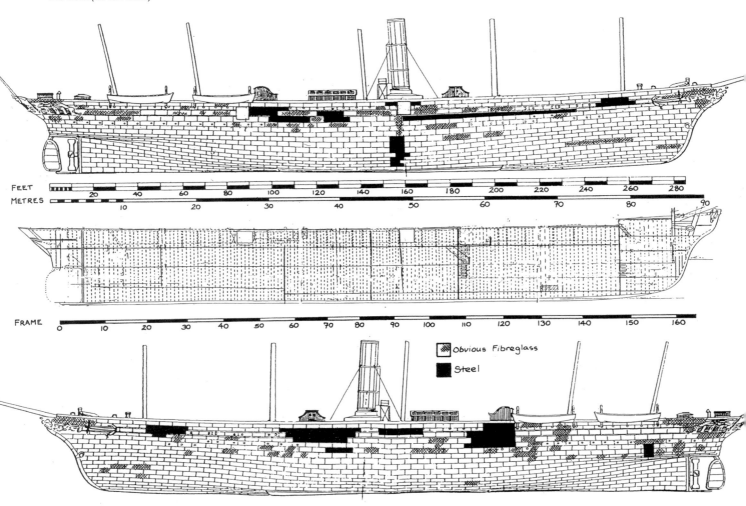

Replica engine

The working of the marine steam engine is probably beyond the comprehension of the great majority of visitors to the *Great Britain*, but it is at the heart of the ship in more senses than one. A steam engine provides a dramatic spectacle, with its great pistons and crankshaft moving constantly. Nothing survived of either of the *Great Britain*'s engines, but both are well represented in drawings and prints of the period, including the Weale prints that provide the plans of the original 1843 engines. However, the plans also contained enough discrepancies to make it hard to create a full-sized replica engine. In particular, the eccentric gear caused some difficulties.

An engine committee began work in 1978, with support from the Institute of Marine Engineers and the Science Museum, among others. It was a long job, but in the meantime the outline of the engine, including pistons, flywheel and chain mechanism, was painted on the after-engine room bulkhead to whet the appetite.

Nearly £750,000 was estimated to be needed, spread over four years, and eventually an Engine Patrons Club was founded in 1993

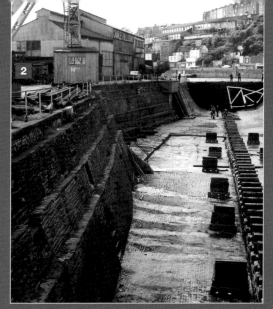

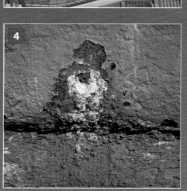

Conserving the *Great Britain*

1 The most fragile parts of the *Great Britain* are those beneath the waterline. They have been enclosed inside a climate-controlled dehumidification chamber, made by roofing the dry dock with glass. The glass 'sea' overhead is both watertight and airtight. *(Jonathan Falconer)*

2 The Great Western Dry Dock in 1970, shortly before the return of the ship from the Falkland Islands. *(ss GB Trust)*

3 Steps in the dockside and the 'altar' edges used in supporting the hull of the ship. *(Jonathan Falconer)*

4 Iron corrodes when it comes into contact with water and air forming a number of substances, the most common of which is rust (chloride infested rust is seen here). The lower part of the *Great Britain*'s hull is worst affected by corrosion because of its exposure to salty seawater over 127 years. Research shows that the iron ship can survive if she is protected from humidity in a very dry environment. At 20% relative humidity (RH) or less this kind of salt-driven corrosion virtually ceases. *(Jonathan Falconer)*

5 There are two giant dehumidification machines that filter the air inside the dry dock and the ship's hull to reduce RH to a constant 20%, which means that corrosion in the hull can no longer continue. This is *Deep Thought II* on the dock side. *(Jonathan Falconer)*

6 Engineer Mick Conway monitors the operation of *Deep Thought I* inside the ship. *(Jonathan Falconer)*

7 The dehumidification machines work by sucking in air and drying it by forcing the air through a water-absorbent chemical powder, before blowing the resulting dry air up over the surface of the ship and through its interiors. Ducts collect the blown air to recycle it through the machine. *(Jonathan Falconer)*

8 The unique display environment for the *Great Britain* impressed not only the HLF, but visitors to the ship as well. *(Pawel Libera)*

to find sponsors. Costs rose with inflation and by 2004 the Trust was looking for £400,000 for a contract to assemble the engine. Finally, with significant support from Rolls-Royce, enough was raised and a detailed specification created to allow a contract to be given to Heritage Engineering of Glasgow. The new 90-tonne engine was given the go-ahead in 2006. It is not 'real' in the sense of being powered by steam. Instead it is driven by electricity, but its pistons and flywheel give a strong impression of how the original would have looked and felt.

ABOVE The outline of the engine painted on to the stern bulkhead as a preview of its reconstruction. *(ss GB Trust)*

BELOW Lowering the new crankshaft into place in 1996. *(ss GB Trust)*

Re-rigging

During the recovery from the Falkland Islands, the ship's three masts, weighing up to 30 tons, were removed with some difficulty. Since the ship was to be restored to her 1845 condition, the six-mast, largely fore-and-aft rig was adopted. Though it had more masts, it was generally simpler than the later square rig, and easier to meet health and safety requirements in that fewer ropes and blocks would be overhanging the public. Nevertheless, there was a near disaster in 1993 when a 6ft (2m) section of the bowsprit fell on to the caisson, fortunately causing no injury.

The re-rigging of the ship would proceed slowly. In 1976 the steeplejack firm of Beaumont's offered money for a new funnel and mainmast, the only square-rigged one. A foremast was fitted, followed by the third mast in 1983 to give something of the appearance of a balanced rig. Two more followed by 1985, made by Harry Spencer at Cowes; but No. 4 remained stored on a lighter because it might have interfered with access for the long-delayed restoration of the engine. In 1997 it was found

to be decayed and was brought to Bristol where it could be looked after. By 2000 the restored 5 and 6 masts were in poor condition and the HLF grant did not allow for their replacement as it was not then considered by HLF part of the ship's structure, or necessary for interpretation. An entirely new rig was constructed from 2005. New steel lower masts were made, though they did not have the taper of the original. Peter Allington of the National Maritime Museum made wooden topmasts and fitted them with standing rigging in 2005, though for the moment there was no money for running rigging. Meanwhile, the ship's foremast, probably dating from the 1850s, had been displayed at the National Maritime Museum in Greenwich for many years. In 1997 it was brought to Bristol to join the mainmast on display alongside the ship. The mizzen remains as a memorial on the front at Port Stanley. The masts and rigging are an essential part of the appearance of the ship. The increasing rake of the masts and the elliptical outline of the mastheads give an elegant and intriguing appearance from a distance.

Interior

The original vision of the ship had always allowed for the passenger spaces on board to be fully restored, and it fitted in well with the vision of the curator in 1998 for 'bringing the ship alive' by recreating the passenger experience – something that every visitor could relate to.

The dining saloon had already been opened for business by 1992 and was used for various social events. It was rebuilt in 2005 and fitted with reproduction tables and benches with reversible backs, as on the original, so that about 96 diners could be seated. A modern kitchen had to be fitted discreetly nearby. Part of the space originally taken up by the boiler room was also used for functions. Replica passenger and crew cabins were fitted, plus a ladies' boudoir funded by a donation. The rest of the forward part of the ship was devoted to displays on the steerage passengers to Australia, ship's galley, stores, toilets etc., soldiers to the Crimean War, and with a part of the hull left exposed to show the structure.

LEFT This is the first reconstructed passenger cabin inside the *Great Britain*, completed in 1997. *(ss GB Trust)*

BELOW Craftsmen apply gold leaf to the dining saloon Corinthian capitals in 2003. *(ss GB Trust)*

Display

A historic ship does not always have a museum attached – it might stand alone, or it might have displays on board, as the *Cutty Sark* did for many years. As a passenger ship, the *Great Britain* allows a great deal of scope for reconstructed cabins and saloons, which take up most of the interior space, so it was considered best that the narrative of the ship should be told separately. The first is an immersive and emotive journey, the second a more familiar intellectual one. A museum was proposed as early as 1982, but there were difficulties over the site and where the Hillhouse Collection of ships' plans – from a shipbuilder working alongside the Floating Harbour at the time of the *Great Britain*'s inception – would be housed; these plans were sold suddenly to the National Maritime Museum instead. Meanwhile, many objects, both large and small, were coming into the possession of the organisation.

The museum and ship experience, designed by the ss Great Britain Trust with David McCabe Design, and opened in its present form in 2005, takes the bold and unusual step of showing the history of the ship backwards, from the recovery to its original conception by Brunel. It gives an opportunity to display many small artefacts and documents, such as letters, diaries and newspapers. It also includes much larger items which cannot be put on the ship for one reason or another, such as the lifting stern frame assembly of 1852 that had survived but lay unattended in the dry dock for nearly 30 years until it was fitted with a new propeller and put on display. There is also an interactive steering wheel display and a replica mast and yardarm among many smaller items, including paintings and domestic goods. The visitor is now ready for their voyage on board. On the ship, as in the museum, visitors have free range over most areas, directed only by signs recommending routes.

As soon as the ship entered the Great Western Dock in 1970 there was a surge of visitors. Although only 20 people were allowed on board at a time for safety reasons, 2,000 a day came aboard during the first few weeks. After the initial interest the *Great Britain* was in for the long haul – the empty hull had few attractions. But visitors are essential for the future of a preserved ship, not just to raise money but also to justify her existence, to raise

RIGHT A view inside the SS *Great Britain* museum on the quayside adjacent to the ship. *(ss GB Trust)*

her profile and to educate a new generation with school parties. The millionth visitor came on board in 1978 and was rewarded with a medal and a glass of champagne. The ship was open to visitors during the International Festival of the Sea in Bristol in 1996, though the event was focused on the replica of John Cabot's ship the *Matthew* of 1493, which was moored just outside the dock. The conservation of the *Great Britain* was largely complete by 2005 and on 19 July she was 're-launched', when the round-the-world yachtsman Sir Robin Knox Johnson poured a bottle of Mumm champagne over the bows, the same brand used by Prince Albert in 1843. As well as the Gulbenkian Prize for Museums and Galleries in 2006, the ship has won 30 national and international awards, and has welcomed a million visitors since 2005.

No preserved ship is ever completely secure and the Trust receives no financial subsidies to carry out any of its work. For many years HMS *Victory* and the *Cutty Sark* were considered model projects, but both have faced severe difficulties in recent times. The *Great Britain* is now thriving and has its place in the heart of the nation and the City of Bristol. It is to be hoped that it will continue to do so.

ABOVE Sir Robin Knox-Johnson re-launches the ship with a bottle of champagne in 2005 after her restoration. *(ss GB Trust)*

LEFT Members of the team, with Matthew Tanner second from the right, celebrate the award of the Gulbenkian Prize in May 2006. It is awarded to one museum or gallery annually for 'innovative and excellent work'. *(ss GB Trust)*

Appendix

Brunel's ss *Great Britain*

Great Western Dockyard
Bristol BS1 6TY
Phone: 0117 926 0680

Brunel's ss *Great Britain* is a fully accessible all-weather attraction.
Tickets allow for FREE unlimited visits for a year and include audio tours.
Visitors are also able to pre-book tickets from the ss *Great Britain* online shop.
http://www.ssgreatbritain.org

Opening times:
Open daily, from 10:00am, except 24 and 25 December. Closed on the first Monday after each New Year Bank Holiday.
Closing times:
4:30pm during GMT.
5:30pm during BST.
Last entry one hour before closing.
There is a pay and display car park next to Brunel's ss *Great Britain* (or nearby on-street parking). Alternatively visitors can travel by train, harbour ferry, bike or bus.
Sat. Nav. Reference: BS1 6TY

The Brunel Institute

The Brunel Institute is a world-class conservation and education centre alongside Brunel's masterpiece, the ss *Great Britain*.

It comprises a conservation suite and archive, major reference library, lecture theatre and seminar rooms, education space, teaching offices and a new ticket and reception area serving the whole site.

The Brunel Institute in collaboration with the University of Bristol safely displays, and makes accessible, the ss Great Britain Trust Collection including over 6,000 maritime books, 2,500 ship plans, 35,000 maritime photographs, and diaries and personal letters relating to passengers and crew of the ss *Great Britain*.

How to access the collection

Access to the Brunel Institute is free of charge. No ticket to Brunel's ss *Great* is required.

To access the collection or use the David MacGregor Library you will need to bring a form of identification which is less than 3 months old, and shows your current address, such as a driving licence or utility bill. Mobile phone bills are not acceptable.

Booking isn't required, but if you are visiting in order to see a particular object or section of the collections, please contact the Curator of Library and Archive in advance by phone on 0117 926 0680 or online at http://www.ssgreatbritain.org/brunelinstitute

Opening times:

Monday	Closed.
Tuesday	10:30am–4:30pm.
Wednesday	10:30am–4:30pm.
Thursday	10:30am–5:30pm.
Friday	10:30am–4:30pm.
Saturday	Open first two Saturdays of each month, 10:30am–4:30pm.
Sunday	Closed.

Bibliography

An exhaustive bibliography for the ship is to be found in Jean Young, *The World's First Great Ocean Liner*, Portishead, 2003

Original reports

Brunel, Isambard, *The Life of Isambard Kingdom Brunel, Civil Engineer*, London, 1870

Brunel, Isambard Kingdom, papers in the Brunel Institute

Claxton, Christopher, *History and Description of the Steam-Ship Great Britain,* New York, 1845

Gibbs, Bright and Co., *The Captain's Instructions for the Management of Our Cadets,* Liverpool, 1854

Great Western Steamship Company, Annual Reports

Guppy, Thomas, *Description of the Great Britain Iron Steamship*, in Institution of Civil Engineers Minutes, 1845, vol. 4, pp.151–85

Hawthorne, Nathaniel, *English Journey*, London, 1941

John Laird, extracts for the Laird Archive, 1840–2, including letters from Dupuy du Lome

Liverpool Compass Committee, *Extracts from 1st and 2nd Reports*, 1856–57

Weale, John, *The Great Britain Steamship of 3500 Tons,* 25 folios of engravings, London, 1847

Periodicals

Chamber's Edinburgh Journal
Downpatrick Recorder
Illustrated London News
Mechanics Magazine
Practical Mechanics

Standard textbooks on shipbuilding, seamanship and engineering

Bourne, John, *A Treatise on the Screw Propeller*, London, 1852

Bourne, John, *A Treatise on the Steam Engine*, London, 1847

Brady, William, *The Kedge Anchor or Young Sailor's Assistant*, New York, 1857

Dana, Richard Henry, *The Seaman's Manual*, London, 1863

Fairbairn, William, *Treatise on Iron Shipbuilding*, London, 1865

Fairbairn, William, *Iron, its History, Properties and Processes of Manufacture*, Edinburgh, 1861

Grantham, John, *Iron Ship-building, with Practical Illustrations*, London, 1868

The Imperial Cyclopaedia of Machinery, Glasgow, 1853

Murray, Robert, *Rudimentary Treatise on Marine Engines*, London, 1852

Reed, Edward, *Shipbuilding in Iron and Steel*, London, 1869

Robinson, Robert Spencer, *The Nautical Steam Engine Explained*, London, 1839

Rosser, W.H., *The Self-Instructor in Navigation and Nautical Astronomy*, London, 1885

Young, Arthur, *Nautical Dictionary*, Dundee, 1855

Local and company history

Albion, Robert Greenhalgh, *The Rise of New York Port 1815–1860*, Newton Abbot, 1970

British Association for the Advancement of Science, *Bristol and its Adjoining Counties*, Bristol, 1955

Buchanan, R.A., and Williams, M., *Brunel's Bristol*, Bristol, 1982

Farr, Grahame, *Shipbuilding in the Port of Bristol*, Greenwich, 1977

Gren, André, *The Foundation of Brunel's Great Western Railway*, Kettering, 2003

Hartree, Richard, *John Penn and Sons of Greenwich*, 2008

Malpas, Peter, and King, Andy, *Bristol's Floating Harbour*, Bristol, 2009

Ritchie-Noakes, Nancy, *Liverpool's Historic Waterfront*, London, 1984

Trinder, Barrie, *The Most Extraordinary District in the World – Ironbridge and Coalbrookdale,* Chichester, 2005

SS *Great Britain* logs

Anon, June 1854–January 1855, National
    Maritime Museum, Greenwich
Anon, March–August 1855, National Archives,
    Kew
Anon, September 1855–June 1856, National
    Maritime Museum, Greenwich
Bright, Reginald, Engineer, August–November
    1852
Cunliffe, Richard, Engineer, July–September 1873
Cunliffe, Richard, Engineer, June–November 1874
Hedges, Phillip, Chief Officer, July–August 1845
Hosken, James, Captain, July–August 1846
Matthews, Barnard, Captain, May–June 1852
Matthews, Barnard, Captain, August–November
    1852, November 1852–April 1853
Matthews, Barnard, Captain, August–December
    1853

Crew lists are available for most periods and are
indexed in the ss Great Britain collection at the
Brunel Institute. There are passenger lists for
nine voyages.

Ship's newspapers

The Albatross, 1863, reprinted nd
The Cabinet, 1862, reprinted nd
The Great Britain Magazine or Weekly Screw, 1861
The Great Britain Times, 1865, reprinted 1975
The Vain Effort, 1860, National Maritime
    Museum, Greenwich

Diaries and letters

Anon, Diary, 1863
Archer, Samuel, Diary, 1857
Aspinall, Clara, *Three Years in Melbourne*,
    London, 1862
Bright, Anna-Maria Georgiana, Letters, 1875
Bright, Heywood, Letters, 1855
Caffyn, William, *Seventy-one Not Out*, London,
    1899
Campbell, John, Diary, 1871–76
Crompton, Mary, *The Journal of a Honeymoon
    Voyage on the SS Great Britain, 1866*, 1992
Dearlove, Mr, Diary, 1862
D'Ouseley, Rosamond, Diary, 1869
Dunn, Thomas M., Journal, 1864–65
Edwards, T.E., *Blue Water Reflections*, in *Sea
    Breezes*, 1922
Fedden, Olcher, Diary, 1852
Gilmour, Allan, Journal, 1852

Grenyer, Felix, *Autobiography*, London, 1923
Gurner, John, *Life's Panorama,* Melbourne, 1930
Henning, Annie and Amy, *Sea Journals,* Sydney,
    1984
Henning, Rachel, *The Letters of Rachel
    Henning*, North Ryde, NSW, 1986
Joseph, Elizabeth, Diary, 1870
Mulquin, Mother Mary, Diary, 1873–74
Parr, Walter G., Account, 1870
Pope, Richard, Diaries, 1867
Saddington, Robert, Journal, 1853
Stap, Henry, Letters, 1882
Towle, Edward, Diary, 1852
Waters, W.D., Log and Diary, 1868

Photocopies of these and many other journals
are to be found in the ss Great Britain collection
at the Brunel Institute.

Modern accounts

Allington, Peter, *The Sailing Rig of the SS
    Great Britain* in *Mariners Mirror*, vol. 84, 1988,
    pp. 444–57
Ball*,* Adrian, *Is Yours an SS Great Britain
    Family?* Emsworth, 1988
Blake, Joe, *Restoring the Great Britain*,
    Bristol, 1989
Corlett, Ewan, *The Iron Ship,* Bradford-on-Avon,
    1975
Fogg, Nicholas, *The Voyages of the Great
    Britain; Life at Sea in the World's First Liner*,
    London, 2002
Gardiner, Robert, and Greenhill, Basil, eds, *The
    Advent of Steam, the Merchant Steamship
    Before 1900,* London, 1993
Goold-Adams, Richard, *The Return of the Great
    Britain,* London, 1976
Greenhill, Basil, and Allington, Peter, *The First
    Atlantic Liners, Seamanship in the Age of
    Paddle Wheel, Sail and Screw*, London, 1997
Griffiths, Dennis, Lambert, Andrew, and Walker,
    Fred, *Brunel's Ships*, London, 1999
McMurray, Campbell, *Ships' Engineers: Their
    Status and Position on Board, c1830–65*, in
    West Country Maritime and Social History,
    no. 13, 1980
Young, Chris, *The Incredible Journey, The SS
    Great Britain Story, 1970–2010*, Bristol, 2010
Young, Jean and Chris, *Monarch of the Ocean,
    an Account of the Great Britain's First Voyage*,
    Bristol, 1995

Index